In a Cold Land

IN A COLD LAND

Saskatchewan's Horticultural Pioneers

by Sara Williams

Grasswood Books

and the

Saskatchewan Perennial Society

Saskatchewan Perennial Society
21024 Grosvenor Park P.O.
Saskatoon, SK S7H 5N9
Canada
website: http://www14.brinkster.com/saskperrennial/
e-mail: spsgarden@hotmail.com

Library and Archives Canada Cataloguing in Publication
Williams, Sara, 1941–
In a cold land : Saskatchewan's horticultural pioneers / by Sara Williams.

Co-published by Saskatchewan Perennial Society.
Includes bibliographical references.
ISBN 0-9780828-0-X

1. Horticulture—Saskatchewan—History. 2. Horticulturists—Saskatchewan—Biography.
I. Saskatchewan Perennial Society II. Title.

SB61.W54 2006 635'.09227124
C2006-902536-3

Editing, design, and layout by Roberta Mitchell Coulter, Saskatoon, Saskatchewan.
Printed in Canada.
Front cover painting: "Lily Garden 12" by Louise Cook of Bert Porter's lily fields at Honeywood. *Back cover photographs:* Percy Wright's 'Thunderchild' rosybloom flowering crabapple; Percy Wright at the Moose Range Nursery, *c.* 1940s, courtesy Ruth Millar.

The publisher acknowledges the financial assistance of the Province of Saskatchewan Community Initiatives Fund towards publication of this book through a Celebrating Community Centennial Grant. Support for the project was also provided by Honeywood (Dr. A. J. Porter) Heritage Nursery, the Saskatchewan Perennial Society, the Plant Science Department of the University of Saskatchewan and the Canadian Prairie Lily Society.

Contents

Acknowledgements

A book like this, small as it is, does not happen without a lot of help. First and foremost I wish to thank Roberta Coulter for her skillful, sensitive, and thoughtful editing, layout, and design. Truly, she could make a silk purse from a sow's ear if put to the test.

For technical reviews of various chapters: Allan Daku, Art Delahey, Darryl Fehr, Rich Krahn, Ruth Millar, Dave Moe, Judy Harley, Peggy Sarjeant, Hugh Skinner, and Koren Vangool.

For financial assistance: Community Initiatives Fund, Celebrating Community Centennial Grant; Honeywood (Dr. A. J. Porter) Heritage Nursery; the Saskatchewan Perennial Society; the Plant Science Department of the University of Saskatchewan; and the Canadian Prairie Lily Society.

For use of photographs and slides: Rosthern Historical Society and the Town of Rosthern, Art Delahey, Judy Harley, Rich Krahn, City of Saskatoon Public Library Local History Room, Friends of the Forestry Farm House, Ruth Millar, Douglas E. Wilde, Joyce Sarsfield, Betty Toop, Hugh Skinner, Brian Baldwin, University of Saskatchewan Archives, Honeywood Heritage Nursery, and the City of Saskatoon Archives.

For the cover painting, "Lily Garden 12": Louise Cook of Saskatoon.

For generous access to papers and help in showing me how to find my way through their archives: Patrick Hayes, University of Saskatchewan Archives; staff of the Local History Room, City of Saskatoon Public Library; Barry Greig, Devonian Botanic Garden Library; Lila Zentner, *The Western Producer* library; Jim Sullivan; Honeywood Heritage Nursery; and Jeffrey O'Brien, City of Saskatoon Archives.

For telling me stories: Allan Daku, Joyce Sarsfield, Art Delahey, John Valleau, Dieter Martin, and Betty Toop.

For proofreading: Paddy Tutty.

I hope I have not left anyone out.

Sara Williams

Introduction

"From an economic perspective, I realize now that Percy [Wright] and Bert Porter, Frank Skinner and so many others followed their passion in life but so far as making a decent income, the time and place was not right. These people must have led an extremely lean existence. How on earth did they ever raise a family?"
John Valleau

Walk down any street or back lane in Saskatchewan and you'll likely come across a 'Kerr' apple, a Patterson lily, a 'Sutherland Golden' elder, a 'Thunderchild' flowering crabapple, or an 'Earlibird' lily. You may not know them all by name, but you'll recognize their flowers, form, or foliage. Many of these plants now grow far beyond our provincial borders, and some have made an enormous impact on the horticultural market worldwide.

While most prairie gardeners are well acquainted with these plants, few of us know of the men who developed them. This book is an attempt to rectify this, to let prairie gardeners and historians know these men as well as the fruits of their labours.

Saskatchewan pioneer horticulturists Cecil Patterson, Percy Wright, Bert Porter, and William Kerr were born far from Saskatchewan around the turn of the last century. To appreciate the full magnitude of their accomplishments, let us go back to the early 1900s and realize that there were no hardy fruit trees on the prairies at that time. In order to get raspberries, it was necessary to bend them down each fall so that the fruiting canes would be protected by snow. Most ornamental trees and shrubs brought from eastern Canada simply died. Yet during this era of prairie settlement, it still behooved pioneer settlers to grow as much of their own fruit as possible.

These men crossed hardy native plants with the higher quality but tender plant material from warmer climates, and bit by bit, through several generations, they came up with hardy, high-quality introductions: plums, apples, strawberries, and raspberries. In keeping with the Biblical verse, they did not live by bread alone. They also introduced beauty into our lives with roses, lilies, perennials, vines, and ornamental trees and shrubs.

While Patterson and Kerr had formal post-graduate horticultural training and the security of public service employment with the University of Saskatchewan or the federal government, Wright and Porter were almost entirely self-taught and led a precarious economic existence in geographically isolated rural areas.

All of these men were contemporaries and knew each other. They were all members of the Western Canadian Society for Horticulture (WCSH), which was established in 1945 and continued until the late 1980s. They met regularly at its meetings and field days, and they were all letter writers. They carried on a prodigious correspondence with each other and other prairie horticulturists. Much of what we know of their horticultural work is through their correspondence, contemporary nursery catalogues, and the annual proceedings of the WCSH.

I was fortunate to have met Percy Wright. He was my neighbour in Sutherland. We met at six o'clock on the morning after we had moved into our house. He was planting a tree in our back yard. Through him, I soon met Bert Porter and enjoyed their friendships for many years. As well, I was given generous access to their papers after their deaths.

I never knew Cecil Patterson or Les Kerr, and it was therefore more difficult to piece together their stories and get a sense of them as persons. Both came to the prairies as adults, and neither have family in Saskatchewan. Patterson emerges as a man absolutely dedicated to his job, an almost living embodiment of the Protestant work ethic, somewhat formal and reserved, with a strict code of conduct and high standards. Les Kerr was perhaps more of a maverick, with a subtle sense of humour and a love of all things wild. In terms of his hybridizing and development of ornamental trees and shrubs, he went well beyond his mandated role.

All of these men shared a total devotion to plants, and we are the more fortunate because of it. Their living legacies will continue not only in the plants already growing in our yards and gardens, but in future plants based on their pioneer breeding work.

This book is dedicated to its subjects. We have derived so much pleasure from their work.

Cecil Patterson (1892–1961)

Cecil Patterson's interest in lilies may have first been tweaked when his car broke down in Prince Albert National Park in the 1930s and he saw a host of native woodland lilies in bloom. By 1950, he had made horticultural history as the first plant breeder to develop hardy lilies in shades of pink and white by making crosses with *Lilium cernuum* and other Asiatic species. Among his selections, 'Edith Cecilia,' 'White Princess,' 'Jasper,' and 'Apricot Glow' are still considered of great value, fifty years after their introduction. 'Lemon Queen' and 'Edith Cecilia' have since been used in the development of thousands of modern lilies.

The first head of the Department of Horticulture at the University of Saskatchewan, his was largely a one-man department for most of his tenure. A quiet man, absolutely dedicated to plants, he worked mostly alone. Originally from Ontario, he quickly saw the need for quality fruit hardy enough to withstand a prairie winter. His legacy includes the 'Prairie' and 'Patterson Pride' plums, and 'John' and 'Thomas' pears.

Cecil Frederick Patterson was born in Watford, Ontario, on December 1, 1892, one of seven children of Frederick and Mary (née Thompson) Patterson. His father was a cheese maker and farmer. He grew up on the family farm in Lambton County.

Cecil finished grade school by the age of ten and high school at thirteen. He continued to help on the family farm for the next eight years: with the hand-milking of Holstein cows and the hand-hoeing of root crops for their feed, as well as many acres of corn. Hard work was a part of his early life, and even then he recognized its value.

"Dr. Patterson has lived among us in his characteristic quiet way. Doing his chosen work with painstaking thoroughness, never seeking publicity, but always firm in his belief that if a piece of work was worth doing, it should be done well and that his greatest reward would be the contribution which he might make toward a happier, more satisfying life for the people of his adopted province."

W. J. Rae

Patterson was honorary president of the graduating class of the College of Agriculture, University of Saskatchewan, c. late 1920s. Photo PH994420 courtesy of the Saskatoon Public Library Local History Room.

He was, recalls his daughter, "interested in horticulture always." At twenty-one, in 1913, he began classes at the Ontario Agricultural College in Guelph. In 1917, he was the horticultural editor for the college's *Ontario Agriculture Review*. The yearbook calls him "...a good fellow, a genial companion, and a true and generous friend." (Vick) He spent the summers at home, farming. In 1918, he graduated with a Bachelor of Science, specializing in horticulture. He was twenty-six. While at college, he had joined the Freemasons, an organization to which he would belong for the rest of his life.

After completing his degree, he decided to go on to graduate studies. He was very academically inclined: "Being unorthodox in my make-up, I decided to take post graduate work... a very unusual step for a graduate in agriculture in those days." (Vick)

Like many young Canadians of that era, Patterson went to the United States for his postgraduate education. At the University of Illinois in Urbana, he received his master's degree in pomology, the science of fruit growing, in 1920. His doctoral work focused on plant physiology, the study of the internal workings of plants, for which he received high honours in 1921. His Ph.D. thesis examined the growth of bean seedlings in relation to humidity and temperature.

Cecil Patterson, c. 1921. Courtesy Mrs. Joyce Sarsfield.

A One-Man Department

Patterson came to the University of Saskatchewan in 1921, just twelve years after it was established, to set up the Department of Horticulture. He was appointed first as a lecturer in horticulture and soon after became head of the Department of Horticulture. He was twenty-eight years old and described by Dean of Agriculture William J. Rutherford as "a young and enthusiastic expert." (Vick)

Responsible for teaching, research, extension, and administration, he was promoted to full professor in 1924. His was a "one-man department" until Margaret Young joined the staff in 1942. She served as secretary, technician, and administrative aide until 1962. In 1944, David R. Robinson was hired to oversee extension, but Patterson continued to contribute to this program.

From 1922 to 1948, Patterson was responsible for all landscaping and its maintenance on campus. The impressive double row of spruce trees lining the road to the president's residence was his design.

Courses offered included general horticulture, protective forestry, fruit growing, floriculture and landscape gardening, vegetable gardening, a seminar, and apiculture. (Long after the bees were no longer needed for research, ten colonies were maintained for teaching purposes and routine pollination of the fruit orchards.) By 1926, the School of Agriculture was offering courses in horticulture and apiculture. The first degree in horticulture was awarded in 1926. By 1956, seventy-nine students had graduated in horticulture.

Because no textbook on fruit existed that was applicable to prairie conditions, Patterson decided to write his own. Unable to find a publisher for such a specialized book during the Depression, Patterson self-published *Hardy Fruits with Special Reference to their Cultivation in Western Canada* in 1936, dedicating it to Dean Rutherford. *Hardy Fruits*, covering both the science and practice of fruit growing, was aimed at other teachers, students, and fruit growers. Patterson saw a fruit orchard as an essential part of every prairie farmstead, and envisioned small commercial orchards supplying fruit to urban populations.

Cecil Patterson, c. 1920.
Courtesy Mrs. Joyce Sarsfield.

Art Delahey was one of Patterson's students in general horticulture in 1951–52. The course, held in the old Header House, attracted about fifteen students, all men. His first and lasting impression of Dr. Patterson was that he was a "nice, friendly, easy-going man. The lectures were very concise with a lot of hands-on material, especially in propagation. He seldom talked about his own research and never used them as examples." Patterson's strength as a teacher was in the way he dealt with the subject matter. "When you walked out of a lecture, you knew what you'd been taught—his lectures were clear and to the point."

Delahey was surprised years later when he discovered the lilies on Patterson's Sutherland property. "Patterson mentioned hybridizing breakthroughs in lilies but the students never saw them as the classes took place in winter. Examples used in class were mostly house plants such as coleus and geraniums. He mentioned a little about apples, but again because of the season, the students were never out of the classroom with Patterson."

Family Man

Soon after his arrival on campus, the tall, red-headed Cecil Patterson met Laura Fraser, a nurse working in Saskatchewan Hall. The courtship flourished on the old university tennis courts. They were married in Montreal on July 27, 1925 and had three children: Fred in 1926, Edith Cecilia (Eve) in 1929, and Joyce in 1930.

Laura Patterson

Art Delahey

Laura Patterson with the 'Laura Patterson' lily, 1971.

Born in 1895 in Virden, Manitoba, Laura Gibson Patterson won the gold medal the year she graduated from nursing at Saskatoon City Hospital.

She and her husband shared many interests. In winter, Saturday night skating at the university rink was a favourite pastime. In that era, even on the ice, neighbours and colleagues addressed each other as Mr. and Mrs. rather than by first names. They enjoyed bridge too, and this comprised much of their social life during the 1930s. Joyce Patterson Sarsfield remembers her mother in a "gorgeous plum coloured velvet evening gown she had sewn, going out for the bridge evening with Pop."

Laura Patterson also shared his interest in home photography. Their young children assisted in developing film and making black-and-white prints, which Laura tinted with oils and watercolours. The subjects were usually horticultural specimens, but the children were frequently included to illustrate the relative size of a tree or plant.

Joyce Patterson Sarsfield describes her mother as "an incredible homemaker" who enjoyed needlework and reading and was involved in various civic organizations. University salaries were cut in half during the Depression, but the situation was worse beyond the campus. "Mom always had a meal and some warm clothing for the young male transients, who were plentiful."

Herb Sunley, later to make a major breakthrough in lily breeding himself, met Laura Patterson in 1964 when the university was selling about twenty lily cultivars. Many more were growing in the very large garden at the Pattersons' home in nearby Sutherland. It was there that Laura Patterson "kindly acquainted" Sunley "with the many hundreds of different seedlings her husband had bred in his home garden in the evenings. Laura gave me good advice on the breeding of lilies, and supplied me with breeding stock at extremely reasonable prices until she sold her home to the Oblate Fathers."

The lily 'Laura Patterson' was registered by the University of Saskatchewan after Dr. Patterson's death. Art Delahey, a former student of Dr. Patterson's, photographed Laura Patterson with her namesake lily in 1971. He found her a "friendly woman, very charming. She knew what she was doing. She wore her rose coloured coat next to the rose lily."

Laura Patterson died in 1973. She is buried beside her husband and her daughter Edith Cecilia in Woodlawn Cemetery in Saskatoon.

Joyce Patterson Sarsfield believes her parents shared an "extremely happy and supportive partnership." She described her father as "a gentle man, unassuming, pleasant, honorable," who believed one's greatest achievement to be "a life well lived." He was intense about whatever endeavour he was engaged in. "Pop's friends were people who were interested in growing plants and all had a valued opinion. Pop had a great joy in living, loved his work, and would respond to questions about taking holidays with a grin and 'every day was like a holiday when working with plants.' He truly felt blessed to be working at his passion." In his leisure time, he enjoyed reading biography, history, the classics, and books of a religious nature. The family listened to current events over a short-wave radio, for which Patterson had built a tall antenna.

Lonely students and friends were frequently invited to dinner. If the amount of food appeared insufficient, the Patterson children understood Pop's quiet "FHB" ("Family, hold back!") message.

The Patterson house on Elliott Street. Courtesy Mrs. Joyce Sarsfield.

The Pattersons first lived at 1115 Elliot Street, just a block away from the university's Memorial Gates. The house was built for them in 1926. Beyond it, looking north, were only fields. They moved to the large lot in Sutherland in 1953 so that Dr. Patterson would have more space for the plants he hoped to continue breeding after his retirement. The lot in Sutherland was virtually a whole city block. After his death, the property was sold to the Oblate Fathers. They cared diligently for the lilies for several years, but no effort was made to select and register worthy varieties from the over two hundred growing there. Art Delahey purchased the lilies from the Oblate Fathers in 1968. Beginning in 1969, the Horticulture Department and members of the Canadian Prairie Lily Society—Delahey among them—took an active role in registering worthy varieties. Eventually, the Oblate Fathers sold the Sutherland property. The area planted with fruit trees became C. F. Patterson Park, while houses were built on the remainder of the property.

Professional and Public Ties

Highly regarded by his colleagues, Patterson was a member of the Western Canadian Society for Horticulture (WCSH) as well as other professional organizations. He appreciated and supported the work of the nurserymen and felt it would be beneficial if they formed an association whose meetings could be concurrent with that of the WCSH, which would be mutually beneficial to both groups. Although he understood the problems faced by nurserymen, horticultural research was his central focus. In a 1959 *Prairie Garden* article, Patterson admonished that "the policy of not recommending a plant because the plant is not readily available, which is far too common, cannot be considered in the

Art Delahey and his fellow horticulture students gained insight into Dr. Patterson's code of courtesy as well as his sense of humour on a very cold day in January of 1952.

About eight of the students (Art among them) dumped a very loud and very large and somewhat pretentious fellow student into one of the fish tanks that were constructed under the central benches of the old greenhouses, thoroughly soaking him, and then leaving him to his own devices to get home in wet clothes on that winter day.

Patterson—always the perfect gentlemen—gave the victim one of his own suits to wear home in the cold weather, but then he did his duty. He asked the perpetrators to come to his office to discuss a "rather unfortunate situation." He gave them an official dressing-down, noting that they could have been prevented from graduating for such boisterous behaviour, but he asked only that they apologize to the victim and to the department. As they were leaving his office, "Dr. C.F.," as he was known, grabbed the last of the students by his shirt sleeve and, with only the slightest twinkle in his eye, said, "Tell me, how did you get that big guy in there?"

best interests of horticulture." Over the years he wrote numerous scientific articles reflecting his own interests and research.

In his 1948 WCSH presidential address, he stressed the need for cooperative research (to avoid costs of duplication), of fiscal responsibility on the part of public institutions, and the need to decide collectively what research priorities would be most beneficial to the people living on the prairies. "As individuals and as public servants we are often faced with the problems of decisions that are of concern either directly or indirectly to large numbers of people," he wrote. If these decisions regarding research priorities were made cooperatively among various organizations and institutions they would be for the greater good. "Regardless of the form it takes, the closest cooperation between all agencies participating in horticultural investigation in Western Canada is essential if substantial and enduring progress is to be made... We cannot continue to travel alone. We must walk together."

He felt that the information presented at WCSH meetings should be extended beyond the membership. During the 1956 meeting, he recommended that libraries and local horticultural societies be allowed to take out associate memberships—which did not happen in the end—and that one hundred extra copies of the 1956 *Historical Report* be printed—which did.

As part of his extension duties, Patterson wrote pamphlets and articles on a wide range of topics, including lawns, peonies, annuals, glads, bulbs, perennials, judging, fruit, and vegetables. He sent out a regular newsletter to interested gardeners, and gave talks and lectures at field days, short courses, and farm women's meetings. His daughter recalls walking over to the university with him on a Sunday afternoon to staple the mimeographed newsletter. He answered letters and phone calls from the public concerning horticulture. For a number of years he even found time to give lectures on horticulture and beekeeping to the students at the nearby School for the Deaf.

Honours

- Agricultural Institute of Canada, charter member
- American Society for the Advancement of Science, fellow (1940)
- Western Canadian Society for Horticulture, charter member, honorary life membership (1960), president (1946–1948)
- Saskatchewan Horticultural Societies Association, honorary life member (1958)
- Agricultural Hall of Fame, inductee (1973)
- North American Lily Society, E. H. Wilson Award (1999)

Cecil Patterson with daughter Eve (Edith Cecilia), 1929. Courtesy Mrs. Joyce Sarsfield.

In 1927 he played a leading role in the formation of the Saskatchewan Horticultural Societies Association (later to become the Saskatchewan Horticultural Association) and was for many years a judge for that organization. In 1945, he helped develop the first Horticultural Zonation map in the *Guide to Farm Practice in Saskatchewan*.

He addressed local groups on a number of topics. At one such talk to the Saskatoon Horticultural Society on hardy bulbs and forcing bulbs, he noticed Les Kerr in the audience and joked, "I heard the other day that he has been grafting apples onto pine trees—and developed pineapples."

In a 1956 newsletter, Patterson advised gardeners to "be sure to include a few evergreens. Evergreens possess refreshing coolness during the summer and friendly warmness during the winter and their presence bridges the gap between one growing season and the next." In a pamphlet entitled "Gardening," he cautioned that "One should not be too ambitious and attempt more than can be accomplished." He reminded recipients, "If you have gardening problems don't hesitate to write to this Department. We shall be glad to help you if we can do so. We have publications on a variety of topics and these we are pleased to send *free* to any person desiring them."

Long before it became fashionable, he argued that mulching was of "great virtue" in conserving moisture and should be especially practised if mulching materials such as lawn clippings were readily available. He warned against the careless use of 2,4-D and the danger of damaging one's own plants and one's neighbours'. "The labor of years may come to an end in a matter of seconds through the ignorance of the user."

Cecil Patterson judging a glad show, 1960. Courtesy Mrs. Joyce Sarsfield.

Plant Introductions

Almost from his arrival at the university, Patterson carried out a wide range of vegetable, annual flower, and perennial cultivar testing. Each year, there would be different cultivars of tomatoes, corn, beans, peas, and other vegetables evaluated. As well, in 1925–26, 170 peony varieties were planted. In 1930, 400 iris varieties were tested. These were in addition to his ongoing breeding and selection of fruit, lilies, and other ornamentals.

As a plant breeder working for a public institution, Patterson held himself to the highest standard: "There is a great temptation on the part of a worker in plant breeding to put forward a new plant when an opportunity to get publicity for this institution and for himself is presented, even though his better judgement is opposed to this step. It is in the best interests of all concerned… to have introduced only those varieties that are definitely superior to the varieties already known and not have variety lists cluttered up with material that should have been discarded on the first round. My feeling is that one should… keep his sights set high, trained not on something … within easy reach and ephemeral but on something necessarily distant… that will serve the generation well." (WCSH, 1954)

He resisted the temptation to release new varieties prior to thorough testing, even when fellow horticulturists pressured him to do so. Yet by 1958, failing health, his own pending retirement, pressure from the university administration, and the coming fiftieth anniversary "Jubilee" year of the university, all worked together to put pressure on him to release the selections he'd been working on for many years. Between 1959 and 1960, he named and released about sixty of his best introductions. Prior to that, he had named and introduced only twenty-three varieties.

As Grant MacEwan wrote in 1970, "Through many of the 40 years in which he headed his department he worked almost alone and his labors were well screened from public view. Consequently, it was not until this unostentatious man's last few years that the public had any idea of the magnitude of his work: 275,000 apple seedlings, 13,000 plum seedlings, 10,000 cherry seedlings, and 4000 pear seedlings."

Lilies

Patterson began breeding lilies that would be hardy in Saskatchewan in about 1934–1935. Most of his breeding work was done in his own home garden and later transferred to the university grounds.

Art Delahey recalls hearing a story from one of the old-timers: "Dr. Patterson was on his way to Waskesiu with his wife in the early 1930s when his car broke down. Beside the road he saw a native Saskatchewan lily. That tweaked him. He wondered if this

Why So Many Introductions?

The whole question of Patterson's sudden introduction of fifty or so fruit cultivars between 1958 and 1960 remains somewhat of a mystery. In an article in the 1959 *Prairie Garden*, he wrote, "It was felt that the University's Jubilee was deserving of special celebration…" and a "decision was reached to name and release a goodly number of fruit seedlings developed in the Department."

It seems oddly out of character for the cautious plant breeder that all indications lead one to believe he was. He set himself high standards and had written on more than one occasion that one should not be pressured by one's institution or the possibility of fame to release plants prematurely before extensive testing. Almost half a century later, one can only speculate, but it seems clear that Patterson was under pressure from the university and that the decision to release for the Jubilee year was not fully his own.

The fact that he had been diagnosed with colon cancer in 1958 and was aware that his own time was limited was undoubtedly also a contributing factor to this massive release. He was a hybridizer who believed in thorough testing. Perhaps he honestly believed that the university plots, which lacked irrigation and offered plants little protection, constituted sufficient testing and there was no need to test them in additional parts of the prairies or under different conditions. Perhaps he had nagging doubts about the attributes of some of these plants but didn't want them consigned to oblivion before this could be determined. He was aware that his successor, Dr. Stu Nelson, was primarily interested in turf and had neither the interest nor the experience to carry on with hardy fruit breeding. Patterson may have felt it was now or never, and his only choice was to at least let the world know by introducing these selections and hope that down the road they could be more thoroughly tested under varying conditions and in other parts of the prairies.

Sadly, few of these fruit trees are available today. Several factors could account for this. Releasing so many trees that were untested by anyone but himself and only in one test orchard in so short a period placed nurserymen in a very uncomfortable position. From an economic perspective, it would have been impossible to propagate and market all of them. In any case, improved transportation and storage methods for store-bought fruit was starting to limit the market for fruit trees. Choices had to be made, though perhaps not always the best ones.

Nor was Patterson there to promote these introductions to the public. With his lilies, he wrote articles for various newspapers, mentioned them in his newsletters, and showed them at horticultural shows as well as field days and lectures. With his untimely death, none of this happened with the fruit introductions.

Although both Percy Wright and Bert Porter evaluated some of these releases, neither was able to evaluate all of them, and there was a time lapse between their release date and when the young plants reached maturity and fruited.

Many of Patterson's selections never reached the nurseries. Had he had the same close relationship with many of the nurserymen as Kerr, perhaps the scenario would have played out differently. For at least a decade, nurserymen would have been more aware of the characteristics of these plants through observation.

Finally, there was the question of water. All of these trees were grown in the university plots without irrigation and at very close spacing. Like other plants, fruit trees will exhibit different characteristics under dryland conditions than under irrigation: quality and size of fruit, form, and disease resistance or susceptibility. These trees were never given that opportunity.

Lily Introductions

The following Patterson lilies were introduced during his lifetime. Unfortunately, several are now "lost," including 'Fairy Pink,' 'Fuchsia Splendor,' 'Indian Lady,' 'Pink Charm,' 'Rose Cup,' 'Rose Queen,' and 'Strawberry Roan.' The remainder are available from the Canadian Prairie Lily Society (www.prairielily.ca) as well as a few specialized lily nurseries.

'Apricot Glow' (1949)—deep apricot, out-facing flowers of up to 18 cm in late June/July. Very vigorous, 1.25 m plants, with up to thirty-five flowers per stem.

'Jasper' (1949)—old rose, down-facing, pendant flowers up to 13 cm across in July; eight to twelve flowers per 1 m stem.

'Burnished Rose' (1950)—vigorous, sturdy, 1.5 m plants; nodding, pendant, strongly scented burnished rose or apricot rose flowers of 10 cm in early July; up to fifteen blooms per stem.

'White Gold' (1950)—the first hardy white lily on the prairies; creamy white, 10 cm, reflexed, pendant flowers; nine to twelve blooms per 0.75–1 m stem. Bert Porter described it as opening "a delicate apricot which quickly bleaches to creamy white."

'Rose Dawn' (1951)—out-facing, 10 cm, deep old rose flowers with a slight orange tint; vigorous; ten to fifteen blooms per 1 m stem in July.

'Edith Cecilia' (1952)—dainty and attractive pink with a slight salmon tint; forty nodding 9 cm flowers per stem open flat in July, becoming reflexed later; overall impression is candelabra-like. Awarded Reginald Cory Memorial Cup by the Royal Horticultural Society of Great Britain, 1955. It is still widely used to develop new pink lilies.

'Lemon Queen' (1952)—deep lemon yellow, long-stemmed, reflexed and pendant 10 cm flowers in July; vigorous; thirty flowers per 1.25 m stem.

'White Princess' (1953)—its flowers are larger than 'White Gold' but its petals are narrower; considered superior to 'White Gold' in that it is more vigorous, has more attractive flowers, and is later blooming; blooms in late July, early August on 1 m stems.

'Rosalind' (1955)—up to thirty down-facing, reflexed and pendant, 10 cm rose pink flowers with a slight tint of apricot on 1 m stems in July.

'Fuchsia Queen' (1957)—unusual in that the out-facing flowers all tend to face in the same direction as if in a spray; up to fifteen deep fuchsia pink flowers of 10–13 cm per 0.5 m stem in late July, early August.

'Orchid Queen' (1957)—honours Queen Elizabeth; the 13 cm flowers resemble orchids, with three merging colour zones: the outer portion dawn-pink, the middle straw yellow, and the centre pale shell-pink; up to twenty flowers per 1 m stem in July.

'Bronze Queen' (1958)—10 cm reflexed and pendulous bronze flowers with metallic sheen; thirty flowers per 1.5–1.8 m stem; vigorous.

'Fuchsia Lady' (1958)—clear fuchsia-pink flowers with trace of spotting; reflexed and pendulous, 12 cm; up to forty flowers per 1.25 m stem.

'Primrose Lady' (1958)—10 cm, pale primrose yellow, reflexed and pendulous flowers; up to thirty flowers per 1.25 m stem; Royal Horticultural Society Certificate of Merit, 1959.

'Red Torch' (1958)—up-facing, 10 cm, red blooms, up to twenty-five per 1 m stem, giving it a torch-like inflorescence; earlier flowering, long blooming variety.

'Tiger Queen' (syn. 'Crimson Queen')* (1959)—honours Queen Elizabeth; up to thirty pendulous, 13 cm, heavy textured crimson flowers per 1.25 m stem in early August; very sturdy.

'Cardinal Beauty' (syn. 'Crimson Beauty')* (1960)—bright crimson out- and down-facing flowers in late June on 0.6–0.75 m stems. Very effective en masse.

The following Patterson lilies were introduced after his death.

'Cohoe' (1969)**—coral at base, red at apex, resembling colour of cohoe salmon, with a few fine spots on lower third of petals; half reflexed, out-facing flowers; 1 m; July.

'Honey Queen' (1969)**—soft yellow flowers with rose tips, lower third slightly spotted; petals half reflexed; 1–1.5 m; July; WCSH Merit Award, 1975.

'Northern Splendor' (1969)—deep rose flowers with blush of yellow, lower two-thirds spotted red-purple, petals half reflexed; short stems.

'Snowdrop' (1969)**—down-facing, pure white, unspotted flowers; yellow-green nectaries; outer petals reflexed; 1 m; July.

'Laura Patterson' (1970)—out-facing to nodding; rose with band of orange above nectary; half covered with purple spots, orange pollen; petals half reflexed; 1–1.5 m; July.

'Northern Princess' (1970)—(syn. 'Mahogany Mauve'); bright red base, deep rose tips; maroon spots on basal half; light green nectary, dark orange pollen; fully reflexed; 0.5 m; July.

'Rose Marie' (1970)—up- and out-facing; crimson throat with light cream tips; pink-purple spots on lower third give illusion that base is darker; deep orange pollen; 0.5–1 m; July.

'Golden Princess' (1974)—down-facing yellow-orange flowers; basal two-thirds has greyed purple spots; nectary green; petals strongly reflexed; 1–1.5 m; July.

'William White' (1974)—up-facing yellow-white flowers fade to white; basal two-thirds is spotted grey-purple; yellow nectary; slightly reflexed petals; 0.5 m; July; honours a former dean of Agriculture at the U. of S.

'Brown Ivory' (1984)—down-facing, amber ivory flowers with brown spots; 1 m; July.

---

*An attempt to register these in 1968 was unsuccessful because someone else had already used the names. They were registered as 'Tiger Queen' and 'Cardinal Beauty.' The dates in brackets indicate their original dates of introduction rather than registration.

** These lilies were introduced from Patterson's own garden (not the university plots) by Art Delahey and other members of the Canadian Prairie Lily Society and the Dept. of Horticulture at the U. of S. after Dr. Patterson's death. According to Delahey, "Dr. Patterson did a lot of his work at home and either intentionally or through oversight that work was not incorporated into his university research."

lily could be crossed with a tiger lily. That's how it all started."

Roger Vick told a different story: "The initial inspiration that launched Patterson on his lily work came from a pioneer Saskatoon physician, Dr. H. D. Weaver, who grew large and varied populations of the nodding lily, *Lilium cernuum* [native to Asia with small violet-mauve flowers]. Patterson was particularly intrigued by one of Weaver's seedlings that produced white flowers with some markings. He had come to believe that the commonly grown orange tiger lilies had lost much of their earlier appeal, and felt that if a variety of colors could somehow be incorporated into hardy lilies, then there would be a resurgence of interest in this flower."

Dr. Patterson named the lily 'Edith Cecilia' after his daughter, who died of leukemia in 1942 at thirteen years of age. As Roger Vick wrote, "The death of Edith was a loss that the family found difficult to reconcile. Dr. Patterson was able to cope by taking on an even heavier work schedule. He also served as church elder for many years, and as superintendent of Knox United Church Sunday School."

As his daughter Joyce Patterson Sarsfield wrote, "The greatest sorrow of my parents' lives was the death of Edith (Eve). Treatment for leukemia was limited to blood transfusions at that time, so our dear friend and neighbour, Grant MacEwan, cancelled his holidays and became Eve's blood donor in the pre–Red Cross transfusion era.

Laura Patterson with Joyce, Eve, and Fred. Courtesy Mrs. Joyce Sarsfield.

Most garden lilies on the prairies were orange in the 1930s, when Patterson began his lily breeding work. He wanted to breed hardy lilies in colours other than orange, especially pink and white, and with trumpet-shaped blooms. He succeeded in the first objective but not in the second, although not for lack of trying. As he admitted in the 1958 *Prairie Garden*, "Many thousands of seed lines have been grown… from the use of a trumpet-flowered lily as one parent but the desired combination appears to have failed to take place."

His 'White Gold' (1950) and 'White Princess' (1953) were the first truly hardy white lilies for the prairies. Many of the resulting hybrids were good plants but sterile and therefore unsuitable for further breeding.

He was cautious about releasing new lilies: "Experience has shown that a period of ten years… is required to convince the introducer that a given selection in very hardy lilies is worthy of introduction and to multiply that selection to the point where introduction is feasible." (NALS Yearbook, 1959)

Among his selections, 'Edith Cecilia,' 'White Princess,' 'Jasper,' and 'Apricot Glow' are still considered of great value today, fifty years after they were first selected and introduced. 'Lemon Queen' and 'Edith Cecilia' were subsequently used as parents in many modern lilies.

His work with lilies continued for more than twenty years. To give an indication of what this entailed, in 1959 alone, when his health had begun to fail, he planted 3,300 hybrid seedlings.

Thirty-four of the Patterson lilies are now registered with the Royal Society of Horticulture in London, England. Some of these were registered after his death by the University of Saskatchewan with the assistance of Mrs. Patterson and members of the Canadian Prairie Lily Society. A few of his earlier introductions had "gone missing" but turn up periodically in private gardens.

As Bert Porter wrote in the 1963 Honeywood catalogue: "These lilies have made horticultural history, not only in this region but throughout the world." W. A. Cumming concurred in 1978: "…His *Lilium cernuum* crosses were one of the most important contributions to hardy lily breeding on the Canadian prairies, in fact even internationally. It is regrettable that Dr. Patterson did not receive the recognition he deserved, but he was quite reserved and very seldom left Saskatoon."

Patterson began with *Lilium cernuum*, a rather dainty little pink lily that was hardy on the prairies but not robust. As J. Paul Stevens wrote, "Dr. Patterson's success in first crossing *L. cernuum* with other Asiatic species proved to be a milestone in lily breeding. A seedling produced in 1937 by crossing *L. davidii wilmottiae* with *L. cernuum* was mauve-pink in color with up to 30 blooms. Although never introduced itself, this seedling was used extensively as a parent in further breeding." Crossed with various species, it produced 'Edith Cecilia,' 'Lemon Queen,' 'Orchid Queen,' 'Pink Charm,' 'Rose Dawn,' 'Rose Queen,' 'Burnished Rose,' 'White Gold,' and 'White Princess.'

"To develop winter hardiness he used the Asian species lilies, *L. davidii wilmottiae*, *L. dauricum*, and *L. tigrinum* as well as the hybrids 'Grace Marshall' and 'Oriole.' He used the relatively weak *L. cernuum* and the tender *L. regale* to impart white."

Vegetables

Patterson began potato breeding in 1930. Due to limited greenhouse space, he was "only" able to grow five thousand to seven thousand new seedlings in pots each year. His object was to develop a potato with smooth skin, shallow eyes, and resistance to common scab. In 1949, he became the first person to produce viable seed from 'Netted Gem.' He crossed it with 'Earliane' in a greenhouse under controlled conditions of temperature and humidity.

One of the resulting seedlings, 'Saskatchewan Russet' ('Netted Gem' x 'Earliane') was released in 1960, six months after Patterson had officially retired. It was considered superior to 'Netted Gem' because of its earlier maturity and lack of bitterness sometimes present in 'Netted Gem.' He realized that the new potato's value might lie in breeding rather than production and hoped that "undue pressure to release new varieties will not be brought to bear on the Department."

According to Don Dabbs, retired professor at the University of Saskatchewan, 'Saskatchewan Russet' was thoroughly tested in the Canada-wide potato trials, but its relatively low yield under field conditions was its undoing. But, as Patterson had predicted, the Canadian potato-breeding program used it in their subsequent breeding.

Work with squash began in 1940, and in 1950 Patterson introduced 'Perfection,' a buttercup type lacking the button or turban. The absence of the button made pathogen entry more difficult, giving it a longer storage life.

In the 1956 newsletter, he wrote: "Our 'Perfection' squash continues to be popular among those growing it. It is a small to medium size squash that is early maturing and of good quality... Seed is available at 25 cents per packet from our Department." 'Perfection' was readily available until recently and is still carried by specialty seed houses.

Fruit

Patterson's major breeding program was the development of improved fruit that would be hardy on the prairies. As he stated in *Hardy Fruits*, fruits that "can't stand the climate without special protection… are not worth growing."

With his advanced degrees, Patterson was well qualified to develop this program. In 1923 he made arrangements to use the orchard at the Rosthern Experimental Station containing Saunder's Siberian crab and apple hybrids as well as pure Siberian crabs as female parents. He pollinated them using selected apple varieties shipped in from the Summerland Station in British Columbia. The seedlings from these crosses were then grown on university land, although cooperation with Rosthern continued until 1932. Most of his breeding was done with controlled crosses with pollen from known sources.

Patterson was only sixty-eight when he died of colon cancer on February 20, 1961. During his nearly forty years at the university, he had taken no sabbatical and virtually no holidays. As Margaret Young remarked, "His work was his pleasure." (Jardine) According to his daughter, "In all the years he never sought any accolades for his accomplishments and requested a simple funeral with no eulogy." Later that year, the heavily wooded, 13-kilometre-long Patterson Peninsula separating Hunter's Bay from Lac la Ronge was named in his honour by the Geographical Board of Canada.

Plums

The following plum hybrids, introduced in 1960 from selections first made in 1942, are still available. All are open pollinated hybrids of the native plum *(Prunus nigra)* and the Asian plum *(Prunus salicina)*.

'Prairie'—4–5 cm, slightly pointed fruit; skin dark red with heavy bluish bloom; flesh yellow, sweet, juicy; semi-freestone; very good fresh eating, canning, freezing, jam, and jelly; early September; upright, spreading tree of 3 m.

'Supreme'—5 cm, round fruit; skin thin, golden yellow blushed with orange; flesh yellow, very juicy, sweet; August; excellent for fresh eating; very good for canning; tree medium, upright, only slightly spreading.

'Patterson Pride'—4.5 cm, slightly flattened fruit; skin bright red with very light white bloom, tender; flesh bright gold, firm, sweet, and juicy; freestone; excellent for fresh eating; very good for freezing and jam; good for canning and jelly; mid-September; semi-dwarf with weeping branches.

'Superb' (syn. 'Perfection')—4–4.5 cm, slightly elongated or flattened fruit; skin is yellow with almost complete red wash; flesh is yellow, sweet, very juicy; clingstone; excellent for fresh eating, good for canning and jam; August; upright spreading tree.

'Geddes'—4–5 cm fruit; skin yellow with slight blush; good for dessert; mid-August; 3 m spread; "worthy of wider use."

Cherry-Plums

Nine cherry-plum selections, first made in 1945–46, were released in 1960. Named after the Greek alphabet, all were crosses of *Prunus besseyi* x *P. salicina* (open-pollinated seedlings of Dr. Hansen's 'Sapa,' 'Oka,' and 'Opata'). Of these, only 'Beta' remains readily available.

'Beta'—2.5–3 cm, round to oval fruit; skin reddish blue to dark purple, thin, tough, and slightly astringent; flesh green, moderately firm; late August; good for fresh eating and jam; excellent for canning; low sprawling bush.

Apples

Patterson raised more than 200,000 apple and crabapple seedlings, seeking hardiness, drought tolerance, size, and quality. Between 1958 and 1960, twenty apples, from crosses made in 1929 and selected in 1945–46, were released from the University of Saskatchewan breeding program. As Percy Wright noted, "At the time that the Patterson apples… were introduced… almost nothing was known about their merits and demerits by anyone but Professor Patterson himself." Only 'Chipman' and 'McLean' are available today. Both were seedlings of 'Columbia' x 'Melba.'

'George B. Chipman'—named to honour the editor of the *Country Guide;* fruit 6 cm; green skin washed with red on the sunny side; flesh creamy white, crisp, juicy, and tart; great for fresh eating, lunch boxes, and excellent for canning whole; stores six weeks; ripens mid-September; tree is round-headed, semi-dwarf, with a good branching habit.

'McLean'—fruit 2.5 in., slightly conical; skin pale yellow with light apricot blush; flesh white, crisp, medium juicy, and sweet; late August/early September, keeps until Christmas; good for fresh eating, cooking, and juice; low spreading tree; branches sometimes weak; fully hardy; resistant to fireblight; fruits annually.

Margaret Young, administrative assistant, secretary, and technician, in the university fruit orchard, c. 1940s. Courtesy Brian Baldwin.

Pears

In a Department of Horticulture newsletter in April 1956, Patterson wrote, "Our Department is glad to report continued progress in its fruit breeding work… We have waited long for these as pear seedlings seldom fruit before reaching the age of 12 or 15 years. All the selections made are potentially new varieties for the west." Of the eight pears released in the Apostle series in 1960, only 'John' and 'Thomas' are commercially available today.

'John' *(Pyrus ussuriensis* x 'Aspa')—fruit 7.5 x 6 cm, pear-shaped; skin thin and yellowing well before the flesh becomes soft; good for fresh eating, jam, and jelly; late September; medium sized, fast growing, round-headed tree to 7 m; flowers white, profuse, and very showy.

'Thomas' (*P. ussuriensis* x 'Bartlett')—fruit 6 cm, round; skin colours before fruit breaks down; flesh firm; a good keeper; late September; good for fresh eating, jam, and jelly; medium sized, fast growing tree to 6 m; flowers profuse, white, and showy; most highly recommended of all the eating pears.

Raspberries

Of the over five thousand seedlings raised and tested in the raspberry breeding program begun in the late 1930s, only two, 'Abundance' and 'Frazer,' were introduced. Neither is sold today.

The university joined the Prairie Cooperative Fruit Breeding Program in 1950, devoting twenty-five acres to the program. Patterson was an ardent supporter of cooperative research. In the 1953 WCSH Proceedings, he reported: "This past year has confirmed once again the belief that this project can be made a great success and can make a real contribution to western horticulture." The project was huge: "5235 apple seedlings were obtained from the Morden station in the spring of 1955, to make a total of 22,500 trees… for the project." The first plums were planted in 1951 and bore fruit in 1955. Raspberry crosses from Ottawa were also included.

Cecil Patterson in potato greenhouse, c. late 1950s. University of Saskatchewan Archives photo A-3263.

All of the fruit tested and developed at the University of Saskatchewan during Patterson's tenure was grown without benefit of irrigation. His daughter recalls "a lot of looking skyward during the 1930s Depression and hoping it would rain." When questioned about the lack of sheltering hedges around the Cooperative Fruit Breeding plot during the course of the 1955 WCSH meeting, he countered that it was a "real virtue in that greater degrees of tenderness and greater degrees of hardiness are thereby revealed," and earlier as well.

In his nearly forty years at the university, he was to introduce a number of plums, cherry-plums, apples, and pears, as well as two raspberries. No releases were made from the breeding program involving apricots, blackberries, currants, gooseberries, highbush cranberries, saskatoons, or strawberries.

Patterson Garden, named to honour Dr. Patterson, was formally dedicated on August 28, 1969. Mrs. Patterson unveiled the plaque. It is an arboretum rather than a garden in the conventional sense—a collection of almost a thousand trees and shrubs labelled with botanical and common names. Open to the public, it serves as a test area for such characteristics as hardiness and disease resistance, as well as a repository of plant material for teaching purposes. Many of the original plantings in this four-acre garden on the corner of College Drive and Preston Avenue in Saskatoon are now majestic specimens.

Other Introductions

Patterson did relatively little work with ornamental trees and shrubs, presumably because much of this was the mandate of the nearby PFRA Nursery Station in Sutherland. Yet he did introduce a rose, a honeysuckle, a flowering crabapple, five gladioli, and two geraniums *(Pelargonium)*. None of these are available today.

Sources

Cumming, W. A., North American Lily Society Yearbook, 1978.

Dabbs, Don, e-mail, 16 Dec. 2005.

Delahey, Art, interview, Dec. 2005.

———, "Hybridizers: Dr. C. F. Patterson," undated.

Department of Horticulture, University of Saskatchewan, "Hardy Lilies," n.d.; "History of the Department of Horticulture, University of Saskatchewan, 1921–1969"; Newsletter No. 11, 16 Apr. 1956.

"Dr. C. F. Patterson," Saskatoon *Star-Phoenix*, 22 Feb. 1961.

Ehman, Amy Jo, "A Reunion of Lilies," *The Western Producer*, 18 Sept. 1977.

Extension Division, University of Saskatchewan, "Gardener's Bulletin," Oct. 1969; *Gardeners' Guild Newsletter*, 27 Dec. 1968.

Honeywood Lilies, catalogue, 1963.

"Hybrid Lilies Registered," *The Commentator*, 3 July 1974.

Jardine, Rose Dulcie, "Forty Years of Service to Prairie Gardeners," *Western Producer*, 25 Feb. 1961.

Krahn, George, Saskatchewan Nursery Trades Association, Landscape Saskatchewan, 1957–1982.
MacEwan, Grant, "Patterson Park," *Calgary Herald*, 8 May 1970.
———, "Fruit Developed on Prairies Making History in Alaska," n.d., n.s.
Patterson, C. F., "Syrups for the autumn feeding of bees," *Scientific Agriculture*, Aug. 1927.
———, *Hardy Fruits*, 1936.
———, "List of Horticultural Plants Recommended for Saskatchewan," SHSA, 1931.
———, "Gardening," n.d.
———, address to the Saskatoon Horticultural Society, 25 Oct. 1945.
Porter, Bert, "Some Early Lily Growers I Have Known," n.d.
Prairie Garden, 1958, 1959, 1960, 1963, 1965, 1969.
Rae, W. J., speech on occasion of Dr. Patterson's retirement, April, 1960.
Robinson, D. R., letter to P. D. Hargrave, 17 Jan. 1956
Sarsfield, Joyce Patterson, letter to author and telephone conversation, December 2005, Jan. 2006.
Saskatoon *Star-Phoenix*, "New Varieties of Pears Offer Hope to Gardeners," 6 Oct. 1958; "Dr. C. F. Patterson, " 22 Feb. 1961; "Garden Dedicated," 28 Aug. 1969.
Sunley, Herb, North American Lily Society Yearbook, 1984
Stevens, J. Paul, "Prairie Profile: Dr. Patterson's Lilies," *Gardener for the Prairies*, Fall/Winter 1995
University of Saskatchewan, news release, 20 Dec. 1960.
Vick, Roger, "Who was Cecil Patterson?" *Kinnikinnick*, Aug. 1992.
Western Canadian Society for Horticulture, Proceedings: 1948, 1950, 1951, 1952, 1954, 1955, 1956, 1960, 1961, 1968, 1969, 1975.
———, "Development of Horticulture on the Canadian Prairies: An Historical Review," 1956, reprinted 1986.
The Western Producer, "Patterson," 18 Sept. 1997.
Williams, Sara, "Prairie Lily Breeders: A Breed Unto Themselves," in Sara Williams, ed. *Hort-Hints III*. Saskatoon: Extension Division, University of Saskatchewan, 1993.
Woodlawn Cemetery, 23 Dec. 2005.
Wright, Percy H., "A Host of Hardy Lilies," *Country Guide*, 1960.
———, "The Patterson Apples," n.d.
———, "Prof. Patterson's 20 New Apples," *Country Guide*, June 1961.
———, personal paper, n.d.
———, "The Achievements of Prof. C. F. Patterson," *The Commentator*, 4 Aug. 1971.
———, "Gifts from the Far North to the North," *Horticulture*, Aug. 1960.
www.prairielily.ca
www.littletreenursery.com

Percy Wright (1898–1989)

Percy Wright did not attend school until the age of fourteen, yet he won the Gold Medal when he graduated from the University of Saskatchewan with great distinction.

He married rather late in life, and strangers often mistook his youngest daughter for his grandchild. Percy took that as a compliment. At sixty-five, he was eligible for both the old age pension and the baby bonus.

A prolific plant breeder, he had no formal horticultural training. His most enduring contributions to our landscape are the 'Thunderchild' rosybloom flowering crabapple and the yellow 'Hazeldean' rose.

He was a prolific writer and mentored a whole generation of gardeners through his weekly articles in the Saskatoon *Star-Phoenix* and numerous others in *The Western Producer*, the *Prairie Garden*, and *Garden Clippings*, as well as British and American publications.

"I was born on July 2, 1898," wrote Percy Wright in his journal, "in a stone house along side the canal at Lachine, Quebec. My father was rector of the little stone church, St. Stephen's Anglican, nearby. He was the Reverend Henry Enoch Wright, born in 1860, and hence 38 years old at the time of my birth. My mother, Dorinda Lavinia Evans of Kingsey, Quebec, was ten years younger than my father."

The family left Quebec in 1907 to homestead near Tramping Lake, Saskatchewan. Percy recalls a nine year old's impression of the land: "We just stopped. It was merely another nameless spot in the interminable plain, and there did not seem to be any more

Percy Wright, Moose Range Nursery, c. 1940s. Courtesy Ruth Millar.

"I am lucky. Here I am at 82, or ten years older than the age at which the average Saskatchewan male dies, and still able to dig and delve in my three nursery plots."

"He did not need to live in a grand house, travel widely or acquire material things. He was satisfied with his family, friends, books, and typewriter and his plants… Percy was a gentleman who lived with dignity in a way that was determined by his own needs and not by the superficial dictates of society. He marched to the beat of his own drum."
Pat Adams, May 1989

reason for stopping there than there had been at any other place. The idea of living there seemed like living on a table, a high plateau exposed to all the winds that blew and right up near the stars.... I set out to see what I could find in this apparently featureless land. The bison herds had crisscrossed it with paths, a deeper one in the centre and a shallow one on each side where the calves trotted at their mother's flanks. Most of them led westward to Tramping Lake."

Like their neighbours, they had soon constructed a sod home. In 1910, Percy's father was appointed Dean of an area that extended from Biggar, Saskatchewan, to Wainwright, Alberta, overseeing twelve churches and a number of young, mainly English student ministers.

His mother was left in her pioneer home with four young children from Saturday morning to Monday evening. "We children would watch for his return, often sound asleep in his sleigh or buggy, while Jack the pony negotiated the home stretch independently."

After a brief stay in British Columbia, where on ten acres the family grew onions which were sold to the government to feed troops, they returned to Saskatchewan. Percy's father became rector of the church at Wilkie while continuing to farm.

Above: Dorina Lavinia Evans Wright and the Reverend Henry Enoch Wright, Percy's parents, date unknown. Courtesy Ruth Millar.
Inset: Percy Wright as a young child, c. 1900. Courtesy Ruth Millar.

A Promising Start

Percy spent the next few years helping on the farm, grounding himself in horticulture, and finishing high school. He attended Normal School in Saskatoon, gaining a teaching certificate in 1921, and teaching in one-room country schoolhouses for the next few years. Teachers were expected to board with local families on a rotating basis. One of his own such experiences was less than ideal: "The Harrises had a large house, so large that they did not attempt to heat it at all. I slept in a room so cold that water froze in it every night, and I had so many heavy blankets on me that I could scarcely turn over in bed."

He continued to spend the summers on his family's farm near Wilkie. "I grew vegetables for sale and sometimes felt that my garden was more profitable to me than my father's farm was to him." In 1922, he was awarded first prize for durum wheat and second prize for field peas at the provincial seed fair.

In his journal, he described his own efforts at dealing with drought in an era when continual cultivation was the norm: "My first resort was... mulching, I had plenty of oat straw... and used it to mulch rhubarb and asparagus, with excellent results... With asparagus, I was careful to... manure first [due to the] high nitrogen need of asparagus. Since manure mulches result in lots of weed seed, which it was impossible to kill before the advent of herbicides, I mulched with straw on top of the manure. It yielded much better when thus mulched and manured."

In the spring of 1925, Wright purchased eighty acres west of Wilkie, growing grain on most of it but reserving five acres for nursery stock, fruit, bedding plants, and vegetables, as well as rhubarb and asparagus. So began the Wilkie Fruit Nursery. He also grew caragana, which he sold to the United States Department of Agriculture's (USDA) shelterbelt program. He obtained 'Pixwell' and 'Abundance' gooseberries from North Dakota, multiplied them, and became one of the first Canadian nurserymen to sell them.

"It would be hard to exaggerate the difference I felt when I became a landed proprietor... and the sense of independence it brought... I enjoyed my years between 1925 and 1929, and a good part of the enjoyment came from the moderate but satisfying success of the Wilkie Fruit Nursery... It endured the years of the Drought and the Depression, and provided me with a good deal of the capital I needed to complete my education."

In the fall of 1925, he began courses in agriculture at the University of Saskatchewan, but switched to arts after two years. "The years 1927 to 1929 found me busy, going home in the summers to manage my nursery at Wilkie, and returning in the winters to attend University.... I passed all my classes that Christmas with As." On May 3, 1929, he graduated from the University of Saskatchewan with great distinction, winning the Gold Medal and the Copeland Scholarship in Arts.

By the early 1930s, he was juggling a number of part-time teaching jobs at the university and had begun a master's program, again spending the summers at Wilkie and the winters in Saskatoon.

Wright obtained his M.A. in English from the University of Saskatchewan in 1931, as well as a five hundred dollar scholarship that enabled him to proceed to the University of Toronto for his Ph.D. Although he did well at the University of Toronto, the drought, the Depression, rising Nazism, and the pending onset of the war in Europe had a profound effect on his psyche.

By the early 1930s, evidence of drought and Depression were all around him. The farms of southern Saskatchewan were becoming depopulated as families moved into the forest belt to try to eke out a living.

"The trekkers had to pass through Wilkie on their way to their refuge [Meadow Lake] across the North Saskatchewan and into bush country. Abandoning the land... they loaded everything ... on their grain racks, hitched a couple of horses to the front, and led another pair, or a cow, behind. On the rack would be crates of old hens, and these would represent the family's only negotiable assets. They would stop at a farm, including our place, and try to trade off these in return for potatoes, bread, or even hay for their horses... It was a most pathetic migration, undertaken in pure desperation."

Solitary, Unsociable, Poorly Adjusted

Chronically ill as a child, Percy Wright did not attend school until he was fourteen, when he entered grade seven, and then only for two years, completing high school through correspondence. He was largely self-educated.

"Why was I not sent to school? I suspect my health was not good enough. On the other hand, I do not recall any schoolhouse in the neighborhood, or even seeing anyone known as 'the teacher'... Whether my father was breaking any regulation in letting me learn at home I do not know, but the type of education which I received was almost surely the ideal. I developed a capacity to learn without being taught, from books and conversations alone...

"I ... spent hundreds of hours on the sofa reading... If I were suddenly dropped into a modern school, with child psychologists and guidance specialists, my report cards would have read: solitary, unsociable, poorly adjusted."

Percy Wright, c. 1920. Courtesy Ruth Millar.

A Gentleman and a Scholar

"My thesis was on Thomas Hardy... his novels, of course, are of uneven quality. Many of them are mere potboilers, although competent ones. The outstanding one is undoubtedly *Tess of the d'Urbervilles* with *The Mayor of Casterbridge* second... Hardy's poetry I did not properly appreciate at the time... the wooden style... as though his metre walked on stilts, distracted me from the meat of his thought... His lyrics... are remarkably un-songlike in form; they have no lilt."

Love at Last

"The summer of 1936 gave me the first glimpse of the girl I was to marry... Ruth Purkis of Toronto, was the driver of one of the Anglican Sunday School by Post vans, touring western Saskatchewan making contacts with youngsters in need of religious instruction. In 1936, the prairies... were desolate in the extreme, with the ravages of drought everywhere evident, abandoned farmsteads every few miles, and those who remained on the land apathetic and discouraged.

"...It was my mother who took a particular fancy to Ruth, and I am sure that the thought crossed her mind that here would be the ideal wife for me—despite the 15 years difference in our ages."

He completed all but his final oral examination, became ill, and never obtained his Ph.D.

He returned to Wilkie, spending the next few years developing his nursery during the summer and working at various jobs for the University of Saskatchewan during the school term. "During the summer of 1938, I did a fair amount of travelling for the Extension Department, to lecture at field days or judge at horticultural fairs. In Nipawin my sister Jennie met me and took me to Moose Range, where she and her husband had set up a general store. I was much struck with the country. It seemed to offer every advantage I needed."

Ruth and Percy Wright with children Ruth, Merry, and David, c. late 1940s. Courtesy Ruth Millar.

He moved to Moose Range in 1939, where he had purchased six acres. "I took with me a large number of hybrid seeds produced by hand-pollination, at least 10,000, all in ink bottles and other glass vessels mixed with sand… One of the new varieties that came out of that seed was my 'Hazeldean' rose, which I still regard as my greatest triumph in plant breeding.

"One year I propagated about 500 crab and apple trees for Prof. C. F. Patterson… all of them new varieties he had developed. The money I earned, at 60 cents per tree, made up a considerable part of my year's income… I continued to propagate crab and apple trees until fireblight, doubtless brought in with package bees from California, got a start in my trees."

He began serious beekeeping soon after the move to Moose Range. "The month of August was unusually warm and nectar kept coming until about September 20. When I gassed off the bees and estimated the amount of honey I had, I was amazed at my success, at least 200 pounds of honey per colony, and all the honey was of the very best quality, mostly from sweet clover." Bees supplied about half his income while he was at Moose Range.

A City Girl

In his journal, Wright often referred to his loneliness as a young man and his despair of ever marrying. "How I envied Frances [his sister], and how I wished that I too could marry…. My own sense of forlornness, and my desperation, were extreme." "I was utterly lonely, and saw no hope of ever having a home of my own, or ever overcoming that feeling of loneliness."

This changed when in Toronto he began seeing Ruth, with whom he had been corresponding. "Ruth and I told each other how unhappy we were, and out of the mutual disclosure came the idea that we could make each other happy… When I became convinced that I could perhaps bring happiness as well as gain it, my hesitation, was overcome." Ruth agreed to marry Wright and came by train from Toronto to Saskatchewan in late July 1940.

"How I remember the late afternoon toward the end of July when Ruth got off the mixed train at the country siding of Moose Range… I did not quite know how to behave." Wright had been boarding with his sister, so with due propriety, Ruth was given the bedroom of the new but empty house Percy had just finished building. No one ever thought that she'd be absolutely terrified to stay alone there, a short walk through the bush from Jennie's but with no plumbing or electricity and the omnipresent fear of bears by the biffy. "Ruth has told me since how primitive she found our conditions… That she overcame her fears and did not back out at the last moment is evidence of a quality she has always had… sheer grit."

As their daughter Ruth Millar wrote many years later, "Dad had spent his earliest years in a sod house in the prairie, while our mother had grown up in a large elegant house in Toronto. What an odd couple they were!" Percy Wright described it thus: "In the spring of 1940, Ruth Purkis, daughter of Thornton Purkis, the advertising magnate who had his offices on Bay Street, Toronto, came out to marry me. At first the loneliness of country life was quite a shock to her, but she soon learned to love the quietness and the freedom, and today wants to move back to a rural living." Ruth Millar is of the opinion that while her mother might have appreciated the quiet and the ambience, she would have preferred it with modern plumbing, central heating, power, and a telephone.

By this point, having purchased new land and built a house, Percy's worldly possessions consisted of twenty-four cases of honey. "If Ruth's father had guessed this fact, he would never have given his permission for our marriage."

"My idea was to take enough of this honey to use on our wedding trip to pay for the expenses. …[A true honeymoon?] So we loaded up what we could in my father's Chevrolet, and headed for Melfort [where they married on August 5, 1940] at which date I was 42, and Ruth, 15 years younger, was 27. She seemed much younger though. She was a mere wisp of a girl then, with a waist so small that she seemed to be willow-like."

Moose Range

All four of their children were born at Moose Range: Ruthie in 1942, Meredith (Merry) in 1944, David in 1946, and Alison Alexandra in 1953. "Our fourth and last child, came so long afterwards that Ruth has called her 'our little afterthought.' I was 55. When a stranger saw me with her, he said, 'Oh, so your grandfather is taking you out today…' I took it as a compliment to my virility." By the time Alison was ten years old, Percy Wright was sixty-five; and the family was collecting both the old age pension and the baby bonus.

Considering her comfortable urban background, Ruth adapted quickly to life in a small rural settlement bereft of electricity and indoor plumbing. The walk to the village to pick up mail or groceries was a few kilometres along the railway tracks with the ever-present possibility of an encounter with bears or timber wolves. She came to treasure the creativity engendered by their summer isolation. As her children grew up, they made up their own plays, songs, and stories and enjoyed the artistic endeavours that came so naturally to their mother. Looking back, she was quietly thankful that her family avoided the pressures of lessons and teams and conformity that living in the city year round would

Sara Williams

Store at Moose Range where the Wrights shopped, 2005.

A fire at Moose Range in the spring of 1944 involving a coal oil lantern proved disastrous, destroying their home and all of their possessions, including Percy's Gold Medal, but fortunately no lives were lost. In the aftermath, "We lived in the honeyhouse for several months, and then put out a part of our $1000 to buy another house, with four more acres of land."

Over thirty years later, Wright asked the university about a possible replacement for the Gold Medal but was told that the die from which the medals had been cast was no longer available.

Hugh Skinner

'Kilwinning' rose

Rose Introductions

Of the over forty hybrid roses introduced by Percy Wright during his working life as a plant breeder, only a very few can still be found in commerce or specialized heritage rose gardens. Among them are:

'Ruth' (1947) ('Mary L. Evans' x 'Alika')—medium red flowers, blooms once, less double but tidier than 'Hansa'; upright pillar type; 2 m.

'Aylsham' (1948) ('Hansa' x *R. nitida)* —bright green glossy foliage, turning bright red in fall; double, deep pink flowers, slightly fragrant; non-recurrent; blooms heavily in June; 1 m; suckers.

'Hazeldean' (1948) *(R. spinosissima* var. *altaica* x 'Harison's Yellow')—better form and scent than 'Persian Yellow'; free of blackspot; sun-resistant, non-fading, semi-double yellow; black hips; fragrant; early summer (June); 2 m. Dr. Patterson sent Wright the pollen of 'Harison's Yellow' from Saskatoon, enabling him to make the cross. Wright felt this rose would prove very useful in current and future breeding, and it was later used in the parentage of 'Morden Sunrise.' John Fopma of Fopma Nurseries, Boskoop, Holland, was given the Canadian rights to 'Hazeldean' for a royalty of 15 percent for three years. 1978 WCSH Merit Award winner.

'Helen Bland' (1950) *(Rosa blanda* x R. 'Betty Bland')—semi-double, slightly fragrant, rose-pink; thornless; upright; vigorous; 2 m; profuse but non-recurrent bloom.

'Yellow Altai' (1950) *(R. spinosissima* var. *altaica* x 'Harison's Yellow')—a sister seedling to 'Hazeldean,' with small, single, bright yellow flowers in clusters; blooms heavily in June; 2 m; hardy, free of blackspot.

'Musician' (1953) ('Hansa' x 'Hazeldean')—semi-double yellow and red bicolour; named 'Musician' for its different tones; hardy, non-recurrent.

'Scarlet Ranger' (c. 1960) *(R. laxa* x *R. gallica)* x ('Gruss an Teplitz')—semi-double bright pink; highly resistant to curculio, blackspot, and mildew; non-recurrent; few prickles; 2 m.

'Poly Won' (1970) (hybrid of 'Gloria Mundi,' a polyantha rose)—hardy, ever-blooming; small, semi-double, pale pink flowers in clusters; no pollen; almost free of thorns; a true dwarf of 30–40 cm.

'Kilwinning' (hybridized in 1948, listed in 1967) *(R. spinosissima* var. *altaica* x 'Harison's Yellow'); same ancestry as 'Hazeldean'—double light yellow flowers, fading to cream; 1.5 m.

Lilies

By the early 1930s, Wright had become interested in lily breeding. He later encouraged both Bert Porter and Herb Sunley to share this interest. In 1968 alone, Wright raised about five hundred lily seedlings. The following are among his lilies still available from specialized nurseries or the Canadian Prairie Lily Society (www.prairielily.ca).

'Nubian' (1950) (seedling of 'Edna Keen')—deep blackish-red down-facing turkscap; the first hardy lily in the dark red classification; Royal Horticultural Society Award of Merit 1950.

'Towering Turk' (1975) ('Red Knight' parentage) —strongly recurved petals, down-facing, deep glossy rich red with darker throat and small black spots; July; 120 cm.

'Home Again'—out-facing seedling of 'Edna Keen.'

'Red Knight' (1957) ('Skyrocket' x 'Sultan') x *L. davidii*—1–1.2 m, down-facing, deep red; 1 m, excellent for massing in the back of border.

'Bold Knight'—hybridized by Wright; named and registered by Herb Sunley; out-facing orange-red; outer petals slightly reflexed.

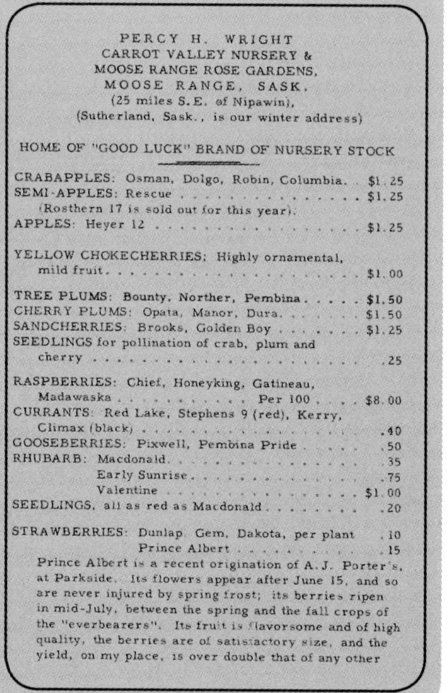

PERCY H. WRIGHT
CARROT VALLEY NURSERY &
MOOSE RANGE ROSE GARDENS,
MOOSE RANGE, SASK.
(25 miles S.E. of Nipawin),
(Sutherland, Sask., is our winter address)

HOME OF "GOOD LUCK" BRAND OF NURSERY STOCK

CRABAPPLES: Osman, Dolgo, Robin, Columbia.	$1.25
SEMI-APPLES: Rescue	$1.25
(Rosthern 17 is sold out for this year).	
APPLES: Heyer 12	$1.25
YELLOW CHOKECHERRIES: Highly ornamental, mild fruit	$1.00
TREE PLUMS: Bounty, Norther, Pembina	$1.50
CHERRY PLUMS: Opata, Manor, Dura	$1.50
SANDCHERRIES: Brooks, Golden Boy	$1.25
SEEDLINGS for pollination of crab, plum and cherry	.25
RASPBERRIES: Chief, Honeyking, Gatineau, Madawaska . . . Per 100	$8.00
CURRANTS: Red Lake, Stephens 9 (red), Kerry, Climax (black)	.40
GOOSEBERRIES: Pixwell, Pembina Pride	.50
RHUBARB: Macdonald	.35
Early Sunrise	.75
Valentine	$1.00
SEEDLINGS, all as red as Macdonald	.20
STRAWBERRIES: Dunlap, Gem, Dakota, per plant	.10
Prince Albert	.15

Prince Albert is a recent origination of A. J. Porter's, at Parkside. Its flowers appear after June 15, and so are never injured by spring frost; its berries ripen in mid-July, between the spring and the fall crops of the "everbearers". Its fruit is flavorsome and of high quality, the berries are of satisfactory size, and the yield, on my place, is over double that of any other have brought. But she would never become accustomed to the intrusion of field mice into her pantry.

Percy Wright, too, enjoyed that time with his children. Their daughter Ruth Millar remembers "the nightly stories he would invent on the spot as we cuddled next to him, all of them cliff-hangers. They were fantasies involving bizarre magical feats. Often, the stories involved flying," something that had fascinated him as a boy on the prairies.

From 1944 to 1954, the Wright family spent the summers at Moose Range and the winters in Saskatoon, where Percy was employed by the Extension Division of the University of Saskatchewan to lecture on horticulture around the province. He also marked papers in English and composition for the School of Agriculture and the College of Agriculture. He looked back on this period as very enjoyable.

"We... raised our own garden... We had our own cow, and a calf which we raised... for meat. We had all the rhubarb, crabapples, plums, raspberries, strawberries, and gooseberries, we could use or preserve. We had unlimited honey, and honey, while sugar was rationed during the war, provided most of our sweets. When we came to the city for the winter... our shipment of goods included as much food as it did household supplies."

The years 1952 to 1954 were extremely wet at Moose Range and much of the nursery stock was lost to flooding, especially the lilies he was in the process of hybridizing. Wright was very discouraged. From 1954 to 1956, Percy again taught at various rural schools, away from home much of the time, and found it altogether a "difficult experience."

A Lasting Legacy

In March 1956 he received a call from Eric Knowles, editor of the Saskatoon *Star-Phoenix*, offering him a job as an editorial writer. He accepted. "I stayed with reporting for quite some time, and enjoyed it immensely. My beat was the university campus, where a number of the professors were the ones I had known when I was a student. I was right at home and found everyone most congenial. I was equipped to deal with abstract subjects and able to quote the professors without error and with the sort of emphasis which they appreciated."

He also wrote extensively for *The Western Producer*, the *Prairie Garden*, *Kinnikinnick*, *Garden Clippings*, and *Horticulture* on a regular basis.

The Wrights bought a fifteen-metre lot at 407 109th Street in Sutherland for one dollar, moved a garage onto it for the first winter, and built first one half of their house and then the other.

"My mainstay during the winter months was the sale of my honey. I would sell what I could to people I knew, and would trade off a further quantity to a general store on Broadway Avenue. For the credits received I drew out the groceries we wanted."

He retired from the *Star-Phoenix* at the end of 1966, at the age of sixty-eight, and began to collect the old age pension of seventy-five dollars per month in January 1967. He taught English at Kelsey Institute for the following two years.

Predeceased by his wife Ruth in 1981, Percy Wright died on April 15, 1989, at the age of ninety after a long illness. His friends and horticultural colleagues Bert Porter and Paul Olsen were honorary pallbearers.

Ruth Millar reflects that her father had given his children many gifts. One was the sense that a wide range of interests could be very enriching, "and one of a myriad of activities is bound to rise to the top, like cream. His second gift, exemplified by the way he had lived his own life was having the courage to do things one's own way, oblivious to peer pressure, norms or traditions. And lastly, he showed his children that we can rise above our infirmities and shortcomings, and create lasting legacies by which people will remember us."

Plant Introductions

By 1979, Wright was eighty-one. He expressed his anxiety about the future in his journal: "I am worried about what is going to happen to my originations when I get older and can no longer do the work that should be done. It would be a pity for them to fail to be exploited and my life work to go to waste."

Part of the problem was that his existing introductions were

Music

Ruth Wright painted this portrait of her husband wearing bifocals at the piano. Courtesy Ruth Millar.

Music was an early and lifelong love. His father had arranged a music correspondence course for Percy from the Columbia Conservatory of Music in Saskatoon in 1914 when the family was still living in Tramping Lake. He spent hours playing the piano, advanced rapidly, and was soon offering piano lessons himself to neighbourhood children as well as playing the piano to accompany silent films at the local hall.

"Teaching music not only provided me with an income... but also gave me a sense of being able to stand on my own feet. Some of the pupils got along well enough to make teaching seem attractive, but others did not... At that time I did not realize just how little musical talent in one's pupils a music teacher was required to contend with." He continued to play the piano until he had to wear bifocals and found it very uncomfortable.

John Valleau recalls visiting Wright in the early 1980s: "I just showed up... He instantly insisted I stay for lunch. There were juice cans on the window-sills, filled with germinating seedlings of who-knows-what, probably roses or maybe fruit trees.

"Somehow we got onto the topic of music and felt some sort of connection. He mentioned he was rather taken with composition in his much younger days... Percy dug through some old boxes and produced a good sized stack of hand-written music manuscripts.

"At this time his eyesight was rather poor so I sat down at the piano and played a few of his compositions. Percy became very quiet and attentive, hearing the notes of his own composing brought back to life after so many years. Then he welled up with tears, and mumbled something in a choked-up voice.

"I left shortly after this, not really understanding that I'd made an old man happy in a very unexpected way. They were surely tears of joy but with a tinge of sadness for things he could no longer manage to do in his sunset days."

not being propagated and sold by nurseries because he was not always taking the time to supply propagating material. Until he was in his eighties, he struggled with the thought that he should stop hybridizing new roses and multiply the ones he had already selected, but he was reluctant to take the time from hybridizing that propagation would require.

He noted, "Simonet... intends to concentrate on multiplying the new roses he has bred up, and to do so will cut down severely on his rose-breeding projects. I feel the same argument, but my impulse is so strong to carry on with rose breeding that I can't see myself eliminating it for several years yet."

Except for a very brief period with 'Thunderchild' flowering crabapple and 'Hazeldean' rose, he made very little money from his hybridizing. As Joe Zary wrote in 1980, "Because Canada has no plant patent law, Wright has not been able to capitalize financially to any marked degree on his originations and introductions."

Other People's Land

Once the family moved to Sutherland in 1956, Wright's plant breeding was limited to his own 15-metre lot. As his daughter Ruth Millar wrote years later, "At home, our city lot resembled a jungle, with no landscaping or apparent order. But many plants in that riot of vegetation were unique."

He soon "borrowed" an acre from Bert Porter at Honeywood (where 'Thunderchild' later originated) and in 1965, a half acre of land at the Veterans' Home on Preston Avenue, and later had some plants growing at George Krahn's Lakeshore Nursery. And he made ample use of whatever space was available on neighbours' lots in Sutherland. It was not unusual to see him planting a tree in the middle of a neighbour's lawn at six o'clock in the morning without prior permission—and getting away with it!

The plot at the Veterans' Home soon became overcrowded. He was not always conscientious about removing poor material. In his journal he wrote, "I am going to have to make a firm resolve to do a great deal more labeling of my plants for it is vexing when one does not know the name of a bunch of young trees... getting names mixed up is my weakness."

In November 1985, Wright sent the following note to the WCSH *Grapevine:* "I am now 87 years old, and my son David says I am "frail," his adjective. I agree that I now fail to do some of the weeding that should be done. My plants are now partly on my home lot here in Sutherland… and [partly] on a farm acreage 15 miles southeast… owned by my MD, Dr. Faddeus Guzik, who has been so favorable to my plant breeding that he charges me no rental and has installed at his own expense a half mile hose 3 feet underground to bring water from the South Saskatchewan River."

Wright, like many of his contemporaries, was a consistent and prodigious correspondent. Over many decades he exchanged letters with fellow plant breeders such as Bert Porter, Robert Simonet, John Wallace, Henry Marshall, and to a lesser extent Cecil Patterson, Isabella Preston, and Felicitas Svejda. His "office" was the chesterfield in the living room. He sat on this with his portable typewriter on his lap. Some of his many papers were "filed" in boxes or piles under it. In his letters, he exchanged propagating and breeding ideas on a wide range of genera as well as pollen, seed, and seedlings.

In addition to the plants listed below, Wright also introduced five other ornamental crabapples, seven honeysuckles, two willows, two applecrabs, a cherry-plum, two pears, a rhubarb, a raspberry, and a strawberry. All of these are today considered obsolete and have disappeared from commerce.

'Thunderchild' Flowering Crabapple

Wright's most famous introduction remains the 'Thunderchild' flowering crabapple, developed on the land that he borrowed from Bert Porter at Honeywood. An open-pollinated seedling of Les Kerr's 'Sutherland,' it has purple-red foliage, bright pink flowers, good form, resistance to fireblight, and a height of 4–4.5 m. It was selected by the Saskatchewan Nursery Trades Association to mark the seventy-fifth anniversary of the Province of Saskatchewan, and in 1971 won a Western Canadian Society for Horticulture (WCSH) Merit Award.

An Eccentric

In the 1970s and 1980s, Sutherland still had a small-town atmosphere even though it had been absorbed into Saskatoon. If a neighbour was somewhat eccentric, you simply accepted it. Wright was one such neighbour. He dressed a bit funny. He wore long underwear year round. You could see it above his collar and sometimes below his trouser cuffs.

Some of the neighbours would swear that when he drove his old station wagon he did so with tunnel vision, never looking to right or left but simply staring straight ahead. Mindful of their young children, they were only too willing to help him put it on blocks each fall.

He often expounded his theories on nutrition and economics, never along orthodox lines. He was the founder of SANE: Society to Aid Nutrition Education. And he was an utter failure at small talk.

As Ruth Millar remembers, "Despite our mother's best efforts to police his sartorial efforts, his get-ups could be picturesque. In winter he wore a beaver hat, and looked professorial. In summer, he wore a battered old felt hat, and baggy pants permanently bulged out from kneeling in the garden."

If he noticed unused space in a neighbour's yard, it made sense to him that it should be planted. He always had hybrid plant material needing a home.

Ruth Millar theorized about the possible origins of her father's eccentricity: "Without the homogenizing influence of school, a child follows his own bent…"

Sara Williams

Percy Wright with one of his own lilies, 1982.

Honours

- Canadian Rose Society, accredited judge, 1972
- Western Canadian Society for Horticulture, honorary life membership, 1978, Award of Merit for 'Hazeldean,' 1978
- Saskatoon Horticultural Society, honorary life membership
- Saskatchewan Horticultural Association, honorary life membership, 1956
- Royal Horticultural Society, England, Certificate of Merit for 'Nubian' lily, fellow
- Saskatchewan Nursery Trades Association, honorary life membership

"If I kept records, I'd never get any work done."

Percy Wright, c. late 1970s

Roses

In his journal, Wright wrote of meeting Isabella Preston, the famous plant breeder from the Central Experimental Farm in Ottawa, in 1936–37. He told her of his ambition to breed superior roses hardy enough for the prairies. "She was rather discouraging and said: 'But you'll have to be a millionaire to do it. You can work and work at rose breeding, and get little or no improvement.'" Fortunately, he didn't take her advice. Unfortunately, he never became a millionaire.

Paul Olsen quotes from a letter Wright wrote to Preston in 1950: "As a matter of fact I am a little proud of my work in roses, not because the results themselves are so remarkable, but because I have done it without any financial assistance, and in spite of extreme lack of funds. Somehow, it just seemed as though there was nothing else to do."

His roses were complex hybrids of hardy species such as *R. acicularis*, *R. arkansana*, *R. blanda*, *R. laxa*, *R. nitida*, *R. rubrifolia*, *R. rugosa*, and *R. spinosissima* var. *altaica* with polyantha or hybrid tea rose hybrids. As Paul Olsen noted, "Percy Wright was a pioneer in using hardy species in rose breeding programs [and] recognized the value of hybridizing the prairie native species."

To give an indication of the magnitude of the work involved, the 1953 WCSH Proceedings reported that he had well over one thousand rose seedlings expected to flower in that year alone and about twenty thousand seeds as the result of controlled crossings done in 1952.

Sources

Adams, Pat, Memorial Address, May 1989.

American Rose Society, *Modern Roses* 6, McFarland Co., Harrisburg, PA, 1965.

Fiala, Fr. John L., *Flowering Crabapples, The Genus Malus* (Portland: Timber Press, 1994).

Friends of the Devonian Botanic Garden, 11, *Edible Apples in Prairie Canada*, September 1991.

Krahn, George, Saskatchewan Nursery Trades Association, Landscape Saskatchewan, 1957–1982.

Millar, Ruth, "My Father's Gifts," Saskatchewan Rose Society Newsletter 7, 3, January 2004.

———, e-mail to author, "Plant Originations made by Percy H. Wright," 21 April 2004.

Olsen, Paul G., "Percy Wright's Roses," Rosebank Letter Rosecom, 25 May 1999, 26 July 1999.

Prairie Garden, 1959.

Robinson, D. R., letter to W. A. Russell, 9 Dec. 1959.

Sunley, Herbert, "Prairie Lily Breeders," North American Lily Society Yearbook, 1994.

Valleau, John, e-mail to author, *c.* 2003.

Western Canadian Society for Horticulture (WCSH), "Development of Horticulture on the Canadian Prairies: An Historical Review," 1956, reprinted 1986.

WCSH *Grapevine*, Nov. 1985.

WCSH Proceedings: 1953, 1963, 1969, 1972, 1978, 1989.

"Percy H. Wright," unsigned typewritten document, 23 Aug. 1976.

Wright, Percy, obituary, Saskatoon *Star-Phoenix*, n.d.

———, conversation with author, *c.* late 1970s.

———, journals (6 vol.), in Devonian Botanic Garden Library.

———, "Pests on Hardy Roses," *Kinnikinnick*, Nov. 1982.

———, "Hazeldean Rose, a Rare Segregation," n.d.

———, "How I Created the Hazeldean Rose," n.d.

Wright, Ruth, conversation with author, *c.* late 1970s.

Zary, Joe, "Plant Patents," 24 May 1980.

Art Delahey

Cecil Patterson's 'Apricot Glow' lily

Art Delahey

Brian Baldwin

Left: *Cecil Patterson's 'Edith Cecilia,' a lily named to honour his daughter who died of leukemia when she was thirteen.* Above: *Patterson was in charge of early plantings at the University of Saskatchewan, including the rows of Colorado spruce leading to the president's residence.* Below: *The old horticulture greenhouses, graced with Patterson's 'Carmine Queen' honeysuckle. Courtesy Brian Baldwin.*

Hugh Skinner

Cecil Patterson's *'White Princess'* lily

Sara Williams

Percy Wright's *'Towering Turk'* lily

Sara Williams

Percy Wright's 'Hazeldean' rose.

Hugh Skinner

Hugh Skinner

Hugh Skinner

More Wright roses. Left: *'Scarlet Ranger'*; top right: *'Victory Year'*; bottom right: *'Aylsham'*

Sara Williams

Bert Porter's 'Redland' lily

Hugh Skinner

Hugh Skinner

Sara Williams

Hugh Skinner

Some of Porter's many lilies.

Left: *'Gale's Favorite'*

Right (top to bottom): *'Cinnamon Toast'*; *'Golden Age'*; *'Happy Thoughts'*

Sara Williams

Sara Williams

Above: *Bert Porter's hazelnut and 'Honeywood' saskatoon*

Below: *Flowering crabapple at the Forestry Farm Park; 'Goldenlocks' elder.*
Bottom: *Kerr's 'Sutherland Golden' elder*

Sara Williams

Sara Williams

Sara Williams

Brian Baldwin

Above: *Les Kerr's 'Plume' caragana*

Bert Porter (1901–2000)

Working alone and in geographic isolation, Bert Porter had neither formal horticultural training nor public or institutional support. Yet he made major contributions to prairie fruit breeding, developing strawberries and raspberries that combined the hardiness of the native fruit he had picked as a child with the quality of tender varieties that wouldn't stand a chance on the Canadian prairies. In lily breeding, he took up where Patterson and others had left off, selecting for better form, purer colours, fragrance, earlier and later flowering, and dwarfness. Although many of his introductions have since been superseded, they have been the basis from which other plant breeders have carried on.

Bert Porter built his log house when he was in his early seventies, using trees from his woods. In his late eighties, he was busy planning for the next nursery season. He read extensively and never owned a television set. While living meagerly himself, he quietly and consistently made an enormous contribution to the life of his community.

Albert John (Bert) Porter was born on April 15, 1901, in Guilford in Surrey, England. He arrived in Canada in 1907 with his mother and three younger siblings—Arthur, Richard (Dick), and Marion—to join their father. John R. C. Porter, a Presbyterian lay preacher and grocery clerk, had begun homesteading the previous year near Parkside, Saskatchewan, fifty kilometres west of Prince Albert.

Bert spent his sixth birthday snowed in at the Regina train station. The family made their way to Prince Albert by horse and

Sara Williams

Bert Porter in Honeywood lily fields, 1985.

"Without formal education and training in the area of horticulture, Bert became recognized throughout the world for his work with lilies and regionally for his work with fruit... Bert's legacy continues to grow as many of his lilies are being used in hybridizing programs throughout the world. Bert himself measured his success by the enjoyment his lilies brought into the lives of gardeners through the years."

Honeywood Heritage Nursery brochure

wagon a few days later, but that spring was the latest the river ever opened. (Williams, 1983) They crossed the North Saskatchewan River near Prince Albert on May 3 with a foot of melt water on the ice. Bert thought it was a terrific adventure. His mother was less impressed and somewhat frightened by the experience. The family lived in a Mr. Bryden's house near Pleasantville before moving to their own homestead near Parkside. They soon acquired two large oxen, some cattle, pigs, and chickens.

John Porter began holding religious services in Shellbrook over a feed barn until a small building and eventually a church were built. He also held services in surrounding communities. His team once went home without him, and he had to walk from Shellbrook to Parkside, arriving at about 3 a.m. (Shellbrook *Chronicle*) He preached in Canwood and Big River for several years, catching the train at Shellbrook on Saturday and returning on Monday.

Bert Porter, Normal School, Saskatoon c. 1923. Courtesy Honeywood Heritage Nursery.

While Bert's father looked after their new community's spiritual needs, his mother, Blanche (née Newman), was soon looking after some of the district's medical needs. The nurse and midwife had no formal training but did have a willingness to come when called. She loved her enormous garden, and the family always had a plentiful supply of fresh vegetables. Through the summer they picked wild fruit—gooseberries, raspberries, highbush cranberries, and chokecherries—for jams, jellies, and canned fruit. Undoubtedly, her love of growing things rubbed off on her eldest son, and his early familiarity with the local native fruit would pique his interest in using it to impart hardiness years later in his fruit breeding work.

His father, who had no previous agricultural experience, farmed a mixed quarter section. Bert and his brother Arthur helped their father clear part of an additional quarter; they left some uncleared for pasture or as bush. This was the quarter section where Bert would eventually live and develop Honeywood Nursery.

The children attended school in an old log schoolhouse in Parkside. Bert spent his free time reading, skating, and swimming. He read the *Children's Encyclopaedia*, a monthly magazine sent to them by an aunt in England. He attended high school at the United Church College in Moose Jaw and Normal School in Saskatoon the first year it opened in 1919. His first introduction to teacher training was only of six weeks' duration, although he would later take the six month "winter course" in 1923–24. In 1923, he worked for a short time with a survey crew at Deschambault and Wapawekka Lakes and later at Flin Flon.

Gentle and somewhat shy, Porter was by nature sympathetic to children, and he took to his chosen profession. He taught in rural schools for the next twelve years, including the school he himself had attended. His first teaching post was at Leney. He was nineteen years old. His students, some in grade nine, were up to sixteen years of age, but Porter was taller. These were the typical one-room schoolhouses of the time, from kindergarten to grade nine, with the teacher boarding with families in the district. The salary was thirty dollars a month.

With the Depression underway, positions became more difficult to obtain. He recalled applying for a job in the early 1930s. "I answered an ad for a teaching position. I got a nice reply thanking me and indicating that I was well qualified. But they said that they wanted a lady teacher as there were so many bachelors in the district, including the members of the school board."

Porter returned to Parkside in 1933 at the age of thirty-two and sold nursery stock for Prairie Nurseries of Estevan for the next two springs. The stock was shipped by railroad, prepaid by the company, with a 40 percent commission going to the salesman. "It was going into the thirties. Everyone was short of money. So sometimes the people who had ordered stuff didn't pick it up. And I ended up with a surplus on my hands." He planted out about thirty surplus apple trees in each of those years on his own farm. Although many of these trees, grafted onto tender root stock, never made it to the following spring, that was the beginning of Honeywood Nursery in 1934.

In the spring of 1934, Bert Porter married Winifred Marshall. Her parents ran the post office in Kilwinning, and her mother was an itinerant nurse and midwife. Their first child, Charles, died at birth. Their adopted son, Richard, later farmed for a few years near Parkside with his wife, Ann, and then went into the gravel business. Their daughter, Joanna Andrews, is a librarian in Montreal.

In 1935, Porter made three hundred dollars from selling ornamentals, fruit stock, and fresh fruit. He continued to substitute teach to supplement his income.

The writings of G. F. Chipman, editor of the *Country Guide*, confirmed Porter's own inclination to specialize in fruit. Porter realized the need for hardy fruit stock on the prairies and concentrated on strawberries, raspberries, apples, and plums for the next fifteen years. His breeding objectives were hardiness, fruit quality, and productivity.

Honeywood's first price list, which was confined to fruit, was issued in 1939. During the 1940s, Porter developed a large market garden at Honeywood, selling fresh raspberries and strawberries. In the spring and fall, he sold raspberry and strawberry plants and in the summer he sold their fruit. Demand for nursery stock was high during the war years as families tried to grow much of their own food to make up for shortages.

Richard and Joanna Porter, c. 1946. Courtesy Honeywood Heritage Nursery.

About this time, Porter was also raising bees for honey production and because the bees

Winnie Porter

Winnie and Bert Porter at the time of their marriage, 1934. Courtesy Honeywood Heritage Nursery.

In 1991, Winnie wrote about their family life at Honeywood: "We were married in Prince Albert in March, 1934. We spent the summer living in a one-room house Bert had built. He was teaching at Honeywood School then, but gave it up after a year to start the nursery, which was very small. We were very poor at that time, but it was in the depression and everyone else was rather hard up too. It was quite a while before things picked up and we made a little to live on.

"In 1943 we adopted a baby boy, Richard Clare, and in 1945 I had a baby girl, Joanna Margaret. They went to school at Honeywood, then when it closed, to Parkside, and then on to high school in Shellbrook. Richard went farming and Joanna went to University in Edmonton.

"We made the house more comfortable and built on a kitchen to convert more room to bedroom space. In 1973 we tore the old part down and built a new house. The old kitchen is now the living room and we have a fully modern homestead instead of just the electricity. We live here all year around. We have no television, but read a lot and play cards. I play records on my stereo and Bert has his radio. I expect we will spend the rest of our lives here."

Winnie forced pots of tulips and daffodils in the old root cellar, entered through a trap door in the living room floor. She intensely disliked lilies, perhaps because they took so much of Bert's attention: "I've had to look at them and hear about them all my life. I've had enough of them." (Brown, 1993) She preferred petunias, pansies, and snapdragons, but she loved roses most of all. "I used to plant seeds and start them in the house. I had nice big windows with good light and grew all kinds of plants from seed. It was a lot of work but I really enjoyed them. Then of course, I had to plant them out in the garden when it got warm enough. Some years that was almost July!" (Cypher, 2005)

She was a wonderful cook, and supper at the big table was always an occasion. One of her favourites was stuffed squash, generally made from Patterson's 'Perfection.'

When John Valleau visited Honeywood as a young man one summer in the early 1980s, he was struck by Winnie's free spirit. "My most vivid memory of the experience had to be Winnie in her halter top, which she wore every single moment that I was there. It shocked me at the time, but thinking back now I figure, good on her! At that age, wear whatever the hell you want. I plan to."

Bert Porter was attuned to nature. He was a shy man, yet when working alone in a lily field far removed from the public eye (as most of them were in the early days of Honeywood), he liked to feel the warmth of the sun on his skin. With gravel roads, dust and noise would herald the approach of visitors in good time for him to don the pair of coveralls hanging on a nearby branch. But with the building of the new highway, things changed. "You can't hear people coming since they put in the pavement," he once remarked. A sign was posted at the entrance of the nursery asking visitors to honk. Some did. Others did not. Surprise encounters with an embarrassed Dr. Porter soon became the stuff of rural legends and probably cramped his style considerably.

Allan Daku tells of the making of one such legend. A neighbour lady stopped at the office in the 1980s to ask for Bert. He had left twenty minutes earlier, telling Allan he was on his way to the asparagus patch to do some hoeing and "get a dose of vitamin D," a code phrase cum warning.

Allan offered to find Bert, but the neighbour insisted that she didn't want to take Allan away from his work. Allan stammered that Bert was getting some sun in the asparagus patch. The neighbour replied, "Oh, I've seen him like that a dozen times if I've seen him once and it don't bother me none."

She walked briskly down the path calling in a loud voice, "Bert, it's me. I need to talk to you, Don't worry 'bout the clothes." A few minutes later she was back in the office, informing Allan that "the ole boy can still run." She winked, got into her car, and drove off. Apparently, Bert was at the north end of the patch and his shorts were at the south end, exactly where the path came out of the trees.

pollinated his fruit. Sugar was rationed during the war, and he sold the surplus honey along with his berries. Although he ceased beekeeping in the 1950s when disease ruined the hives, his interest in them continued, and he was a keen observer. In the 1972 lily hybridizers' "Round Robin," he wrote, "I think they do very little pollinating of the lilies, tho they will feed on the nectar in the blossoms. The chief pollinators of lilies are the hawk moths. They also feed on the nectar and in so doing their wings contact the anthers and so transfer pollen."

Through half a century, long after most nurseries had at least partially automated watering systems, Honeywood Nursery would operate without irrigation, as it still does due to the limited capacity of the well. The nursery soon comprised about twenty acres, during the 1980s increasing to about seventy acres, most of it taken from cropland his father first homesteaded at the beginning of the century. Other land was rented out to local farmers to supplement income from the nursery. Then and now, nursery operations make intensive rather than extensive use of land.

On January 20, 1961, Bert wrote the following letter to the Canadian Wheat Board. It speaks of a time before roads were paved (or regularly ploughed in the winter) and when even a few dollars of extra income was meaningful.

Dear Sirs

I recently received a barley payment cheque for $13.03. On the way home from the Post Office I was stalled in a snow bank and had to shovel out. On arriving home found I had lost the cheque. Went back but could not find it in the snow. Would it be possible for you to cancel this cheque and issue another? The permit book is in the name of my renter, as above.

Yours sincerely

Bert Porter

In 1954, the economic situation of the nursery was still tenuous. In a letter written to Jean Erickson, a fellow lily breeder, Bert wrote: "Have been making an extra big effort this year to

Bert, Winnie, and Richard Porter at *the cabin at Emerald Lake, 1944. Courtesy Betty Toop.*

"Honeywood Nursery and Honeywood Lilies operated for 65 years and was known for a wide range of quality hardy material. Dr. Porter introduced many lilies of his own creation as well as some developed by other hybridizers. Three of Porter's favourite lily introductions were 'Earlibird,' 'Happy Thoughts,' and 'Flaming Giant.' 'Flaming Giant' is known as 'Moulin Rouge' in Holland and is grown by the hundreds of acres for the cut flower trade."

Honeywood Heritage Nursery brochure

get enough stock in enough volume to get us out of debt… My wife is worrying me to give up the nursery business and go back to teaching school. However, I feel that I am doing work here that no one else is doing and it is more important than the money. The question arises, tho, have I the right to expect the family to go along on a lower standard of living than they otherwise would have to enable me to do this?"

In the same letter he wrote that Percy Wright was having an equally tough time: "His nursery was flooded out, much stock killed and more dying now as a result. He writes he has lost at least a thousand dollars worth of lilies. Don't know if he was able to get all of his orders out or not."

From early spring to fall, Porter worked from dawn to dusk. But he always had time to show visitors around and was a gracious host. In the "Round Robin" of December 3, 1972, he wrote, "You were just the beginning of a stream of visitors. In fact, for about two weeks [during the lily season] I was able to do little else but show visitors around. It is always a pleasure but it is time consuming."

Almost from the beginning, Honeywood was predominantly a mail-order nursery business. During the 1930s and 1940s, the mails were more reliable than today. Porter remembered receiving a letter in Parkside at eleven o'clock one morning that had been mailed from Ottawa at 3:30 p.m. the afternoon before.

In 1972, the mail-order part of the business was reduced due to higher postage and labour costs and more limited rail service. A note in the 1972 catalogue read: "It is becoming more and more difficult to carry on a mail order nursery business. Thus we have decided to sharply reduce this part of our business, and plan to confine our mail orders to lily bulbs, peony roots and possibly other herbaceous perennials that can be forwarded by mail."

In the 1972 "Round Robin," he wrote, "I should be making some arrangements for the future as I can't expect to keep on indefinitely. Luckily my health is good. Doc tells me I have the blood pressure of a teenager. Even so, life is uncertain and one never knows at my age what some organ or other may give out. Maybe I'll be like the old one horse shay and go to pieces all at once."

A visitor in the late 1970s described him thus: "Still wonderfully active at 76, Bert Porter's upright carriage, spry step and delightful sense of humor belie his age… a Canadian who dedicated his life to developing better plants for his fellow gardeners and who has enriched gardens around the world." (Stupple)

Allan Daku, a graduate in horticulture from the University of Saskatchewan, joined Porter in July 1978 and became a partner in 1980. By the mid-1980s, under Daku's influence, Honeywood had expanded into perennial production.

Porter was not a man to retire easily, even in his eighties. "I can't see quitting as long as I'm healthy. I will have to make some sort of arrangement, but I'll be working for a little while yet." (Russell) In 1984, a heavy outbreak of botrytis hampered lily production. This was compounded by several years of late spring frosts, which further damaged the lilies. It proved a difficult period. Yet, in 1990, approaching his eighty-ninth birthday, Porter was "busy planning another season at the nursery, deliberately planning for another generation and enjoying the beauty of nature as each day presents it." (Daku, 1991) Later that same year, Bert and Winnie left the nursery, moving to seniors' housing in Parkside.

Bert Porter died at the age of ninety-nine on August 2, 2000, at Whispering Pine Place in Canwood. He is buried in Honeywood Cemetery. As Bev Remus remarked, he was "a kind, quiet, unassuming, tolerant and humble person with a love and total devotion to plants." (Kadota)

Fruit Breeding

In 1934, apple trees sold for about seventy cents each. Among the standard varieties were 'Osman,' 'Dolgo,' and later 'Heyer 12.' Many of the apple trees carried by prairie nurseries had been imported from British Columbia on tender root stock incapable of surviving a cold winter with little snow cover. Most died. Porter began budding and grafting these same varieties onto hardy local root stock. With no horticultural training, he just "went at it."

Some of his earliest hardy stock came from Joe Richards, "an old chap that was growing some fruit a few miles from us. He had a number of crabapples," as well as Dr. Niels Hansen's cherry-plums from the Experimental Station at Brooking, South Dakota, and 'Assiniboine,' a large red plum with yellow flesh. "That was one of the first plums we had that was hardy with us." (Williams, 1983)

Other early stock came from Seager Wheeler of Rosthern. "He was quite a fruit grower. We'd go over there sometimes in the fall and buy fruit from him and I'd plant the seeds of that. You bought the fruit. It didn't matter what you did with the seeds.... So I started growing seedlings and grafting onto them and getting a few fruit trees and increasing raspberry canes and building up a fruit stock." (Williams, 1983) He began in 1938 with a small mailing list, and his clientele grew by word of mouth and reputation.

By 1982, Porter had tried "fifty or more apples" and concluded "not too many are left with us." (WCSH Proceedings, 1982) Many of those that had proved hardy were lost to fireblight in the 1940s. Over the next fifty years, Porter was to breed, select, and introduce many hardy fruit varieties. Beginning with strawberries and raspberries, his introductions also included currants, gooseberries, plums, apples, a saskatoon, a sandcherry, rhubarb, and hazelnuts.

The market garden, concentrating on raspberries and strawberries, paid the bills.

Strawberries

When Porter began breeding strawberries, the standard variety, 'Gem,' was not sufficiently hardy. Beginning in 1943, more than ninety strawberry varieties were trialled at Parkside and thousands of seedlings raised, from both open-pollinated and controlled crosses. Only a few were hardy, and in the end, many more were thrown out because of poor quality or susceptibility to disease.

He introduced nine hardy strawberry cultivars in the next few decades, imparting hardiness by crossing 'Gem' with the local native strawberry. 'Jubilee' and 'Porter's Pride' are still available.

In summing up his strawberry breeding work in the 1960 WCSH Proceedings, Porter wrote: "The above represents the growing of several thousand seedlings. Due to lack of time and help it was not possible to keep records as to the proportions of everbearers to summer bearers, number discarded for susceptibility to mildew, etc. Such records would be of value for planning further breeding work.

"The increase in the number of farm gardens where water is available, plus the great increase in the number of town and city gardens which can be watered, has pointed up sharply the need for strawberry varieties of good size and quality with greater winter hardiness than any we now have."

Raspberries

Among the standard raspberry varieties in the early days of Honeywood were 'Herbert' (with very prickly canes) and 'Viking.' 'Viking' was of very good quality but lacked hardiness. "We'd have to lay the canes down in the fall so the snow would cover them to get fruit." (Williams, 1983) In his breeding program, Porter used pollen from the local wild raspberries he'd picked as a child to impart hardiness and 'Viking' to impart quality fruit.

'Honeyking' (1933) ('Viking' x *Rubus idaeus* var. *strigosus*)—medium berries, very good for canning; good for fresh eating; later used in Agriculture Canada's Morden breeding program.

'Redman' (open-pollinated seedling of 'Honeyking')—fruit resembles 'Viking' but not quite as large; excellent quality; exceptionally hardy; canes up to 2 m.

'Honeywood Black' (1934)—no longer available.

Currants and Gooseberries

Introduced through Honeywood Nursery, 'Willoughby' was obtained from and named after Walter Willoughby of the Melfort demonstration substation located north of Parkside. "They would try and find a farmer who was... interested in horticulture and gardening and supply him with the materials... Once a summer they would hold a demonstration day for the local people to come and see what could be grown. Mr. Willoughby had been supplied with black currants, supposedly 'Boskoop Giants,' but there was one currant that was different from the rest. It was more fruitful for one thing and it was free from mildew." (Williams, 1983)

Memories of Porter's Place in the Forties

by Betty Toop

Every member of our family worked in the Porter nursery in by-gone years. I became a full-time summer worker when Joanna was a tiny baby and that continued for four summers. I was always treated like family.

I enjoyed working in the outdoors picking fruit, hoeing weeds, starting new plants, extracting honey and packing mail-order plants.

The main moneymakers in those days were 'Viking' raspberries and 'Sparta' strawberries. During the height of the season it was common to see as many as twenty workers in the raspberry patch filling the twenty-four pint cases which sold for four dollars (rising to six dollars in my final year). Bert was kept busy moving the crates to a cool spot until they could be marketed. He had an outdoor cold cellar—an underground mud hut where they were temporarily stored.

Betty Clarke Toop, c. 1940s. Courtesy Betty Toop.

Bert was very patient but demanded a high level of performance. This included being careful not to break the tender canes, not to miss any raspberries, to cull out any defective ones, and to top the pint baskets well so the customers got good value for their money. The pickers learned to be skillful, and if instructions were not adhered to they were not asked to return.

The worst menace to the strawberry patch was the lowly robin. One robin could destroy countless berries. They had a habit of selecting the finest as they ran up and down the row, taking a bite out of each berry. These had to be set aside as culls and ended up in Winnie's jam jars.

Outdoor work started as soon as the dew was off the grass and continued, with noon breaks, until 6 p.m.

There was always salad with supper and lots of corn, tomatoes and cucumbers. Winnie made wonderful brown bread and it was my job to cut the slices at the table as required. Much fun was made of my doorstep slices but the bread soon disappeared despite my unskilled efforts.

The evening was always story time for Joanna and Richard and they were brought up on the great old classics like *Wind in the Willows*. This was Bert's quality time with his youngsters. Winnie did embroidery work or made good use of her sewing machine producing her own and Joanna's dresses with great skill.

It was also Winnie's job to name the many varieties of plants and lilies that Bert produced. She had a real knack for coming out with appropriate names. Winnie had a beautiful flower garden under her kitchen window in which she took great pride. Many of the flowers that adorned her garden had been painstakingly started in February in little Avon jars.

She subscribed to McCall's magazine and searched the pages for delectable recipes and the latest fashions in patterns. The tempting cakes, which sometimes stretched her limited budget, were saved for special occasions. I remember being asked over each Christmas season to enjoy the homemade goodies which she made with such care.

I have many fond memories of Joanna and Richard as small children. I especially remember how they loved the rain. They had yellow rain slickers and fireman hats which they donned along with their galoshes. The harder the rain fell, the more they loved it. They splashed happily in the pouring rain belting out the strains of "Jimmy crack corn and I don't care" and as the rain softened so did their tune.

During the years I worked for Bert he was able to buy his first tractor. This enabled him to save countless hours he would have spent hoeing by hand.

When I first worked at Porter's farm, a small three-roomed log building was their home. By 1945 when I worked full time, a large kitchen had been added. Winnie loved the extra space and the former kitchen became a second bedroom. My bedroom was a little nook above the kitchen accessible by a ladder. In the winter, Bert used this space as an office. It was here he prepared his spring orders for plants and shrubs.

Bert never became prosperous but he had the courage to do the work he loved. He was always able to rise above the trials nature brought his way and carry on in spite of his setbacks.

Bert Porter's Instructions for Pickers

- Pick only on the tree or trees assigned to you. If you need more than this—ask where you can get them.
- Remember, apples are as easily damaged as eggs. Place them carefully in your container. Do not drop them unless you like bruised ones.
- To pick, hold the branch by one hand, grasp the apple with the other and gently pull toward the base of the branch.
- To avoid bruising, hang container at your waist or place it on ladder.
- When using ladder, make sure it is on good footing before mounting.
- Do not over-reach when on ladder. Move ladder to a more convenient place.
- Pick your own tree or portion of it clean.
- If you are able-bodied, pick the higher fruit, leaving the lower for older or handicapped pickers.
- Do not allow children to run here and there sampling or throwing fruit. This last rule is precautionary. Most children that come here are well behaved.

'Parkside' gooseberry (1956)—fruit larger than 'Pixwell,' green with a blush on one side; delicious off bush.

'Honeywood Red' currant—heavy fruiting; large berries in large clusters, easily picked. The 'Parkside' gooseberry and 'Honeywood Red' currant should be available from Honeywood within a few years.

'Willoughby' black currant (1935)—fruit of good size, mild flavour; free of mildew. Seventy years after its introduction, 'Willoughby' is still widely sold.

Apples

Porter introduced 'Redheart' crabapple and 'Mystery' apple. Both have been superseded by superior varieties.

Other Fruit

'Honeywood' saskatoon (seedling of a large fruited wild saskatoon growing south of the nursery; selected, 1955; introduced by Lakeshore Nursery, 1973) —fruit 1.5 cm with clusters of up to fifteen berries; mild flavour and good productivity; blooms a few days later than many cultivars, therefore avoiding frost damage; 2–3 m high x 3 m wide; fruits at early age; berries ripen over long period; few suckers. Still available.

'Honeywood' sandcherry (1935)—large-fruited mildew-free sandcherry.

'HoneyRed' rhubarb ('McDonald' x 'Early Sunrise')—with adequate moisture, stalks remain tender until late summer; finely grained, bright red when cooked; non-stringy; requires less sugar; similar to 'McDonald' but with almost no seed stalks. Still available.

Hazelnut hybrids (*Corylus cornuta* x *C. avellana*)—good quality hazelnuts; non-spiny cups easily opened once the nuts are ripe; hybrid of the native hazelnut and European filbert; remarkably productive; seedlings appear to breed true; sturdy upright bushes, 2.5 m; need two plants for fruit set. Still available.

Plums

Porter received his first plum seeds from George Chipman, editor of the *Country Guide* and an ardent proponent of fruit growing. None of these proved hardy. He also raised seedlings of 'Dandy,' which he obtained from Boughen Nurseries of Valley River, Manitoba, and seed of *Prunus salicina* from Norman Mountain of Lloydminster.

In the 1970 WCSH Proceedings, Porter discussed future breeding programs for hardy plums. "I would hope that someone with more years ahead of them than I have would follow this up.

Honeywood Fruit Farm and Nursery

A. J. PORTER, Proprietor, Parkside, Sask.

PRICE LIST: SPRING 1939

I am offering this spring the following nursery stock, all grown here at Parkside. The tree fruits are all grafted on hardy stocks grown from seeds raised here. All varieties listed have produced satisfactory crops of fruit in this district.

STRAWBERRIES

We specialize in strawberries and have tried out many varieties. Gem is the best all round sort we have yet discovered. Dakota is the hardiest but not firm enough for a market berry. Dunlop has done well in many districts. Parson's Beauty, Glen Mary and Dorset are good quality berries that are worth a wider trial. Aberdeen has done well at the University at Saskatoon.

	25	50	100	500	
Gem Everbearing	$ .90	$1.50	$2.50	$10.00	Prepaid
Senator Dunlop	.50	.75	1.00	4.00	Prepaid
Dakota	.50	.75	1.00	4.00	Prepaid
Parson's Beauty, Glen Mary Dorset or Aberdeen for trial	.90	$1.50			

RASPBERRIES

Viking has proved to be the best in quality and the most profitable raspberry we have tried here. Like all the high quality raspberries the canes require winter protection. We can also supply a limited quantity of Newman, Chief or Herbert at the same price.

20 for $1.00 $4.00 for 100 Prepaid

GOOSEBERRIES

Pixwell is the only gooseberry we can fully recommend at the present time. We have a number of new sorts on trial but it is too soon yet to determine their value.

Pixwell Gooseberry	30c each	$3.00 per dozen	Prepaid
Well Rooted Layers	15c each	$1.50 per dozen	Prepaid

CRAB APPLES

The following crabs have been widely planted within the last few years: Osman, Olga, Florence, Dolga and Red Siberian. Prices: 60c each. 3 for $1.65 $6.00 per dozen

These have larger and better quality fruit than the above but are not as hardy and should only be planted in a well protected location. Hayer No. 12, Rosilda, Jacques and Rosthern No. 7.

80c each 3 for $2.25

PLUMS

Assiniboine, Dandy and Olson are three of the hardiest plums we have. Assiniboine is the largest but the other two are firmer and better for canning.

70c each 3 for $2.00

Tecumseh, Ojibwa, Mina: These are hybrid plums and of higher quality than the preceeding. Tecumseh and Ojibwa are less hardy than the natives but generally fruit heavily below the snowline. They should be grown only in bush form. Mina appears to be fully hardy here and is an exceptionally good cooking plum.

$1.00 each. 3 for $2.75 Quantity limited.

PLUM-CHERRY HYBRIDS

These should be grown in bush form only as they frequently kill back above the snowline. However, they produce large quantities of fruit near the ground and should be planted in every farm orchard. Mordena is hardier and produces fruit to the top but requires another late blooming sort such as Ruby, Mansan, or Compass to pollenize it.

Tom Thumb, Brooks or Champa	50c each.	$5.00 per dozen
Opata or Sapa	60c each.	$6.00 per dozen

The rebuilding of the Porter home was a milestone in their domestic life. In 1972, "with a little help from the neighbors to get the roof on" (Russell, 1977), Bert built the log house that is still standing on the property. He was seventy-one years old. He discussed his construction plans at length in the winter 1972 North American Lily Society hybridizers' "Round Robin": "Will have the basement dug out and rebuild somewhat larger. Got about half the needed logs cut and hauled out to a spot where we can peel them in late spring to let them season. It will be rather hectic living in the kitchen part while the tearing down and rebuilding goes on. ... Have been finding some nice straight logs in our own woods here. I bought a power saw this fall for this work. This will make it much less labour than when we built originally. Also, I think we can borrow a tractor with a front end loader to lift the logs as we go up. Will lay this glass wool insulation between the logs to stop any cracks, so should be easier to get a tight wall. Maybe we can arrange for a raising bee when the time comes to put them up. This was how most of the old log house was built, also barns when we first came here—with the "bee" I mean. The tractor and glass wool and power saw are modern conveniences." (Photo courtesy Honeywood Heritage Nursery.)

I would suggest using 'Ivanovka,' the largest fruited of the hardy *P. salicina* plums, as the seed parent, and use on it pollen of the best sandcherry hybrids we now have such as 'Opata,' 'Manor,' or 'Gamma.' This program should be carried through several generations.

"I think we have some valuable character[istic]s in the sandcherry which would greatly improve our plums if they were incorporated into them: later blooming, resistance of fruit buds and blossoms to frost, annual bearing, small pits, retention of the ripe fruit on the tree. What needs to be bred out of our sandcherry-plum hybrids is prostrate habit and susceptibility to mildew on foliage and fruit."

Two of his earlier plum introductions were 'Parkside' and 'Mistawasis.' Both were *Prunus salicina* hybrids that have since been superseded.

By the late 1970s and 1980s, Porter had selected some of his seedling plums as future introductions. The fruit of many of these were entered into shows and won ribbons. Unfortunately, these were not introduced prior to his retirement.

Flowering Crabapple

'Spring Snow' (1967) (Inter-State Nurseries, Hamburg, Iowa)—(8 x 6.5 m); rounded tree form; bright green leaves; single white flowers, no fruit. Available.

This beautiful white-flowering crabapple originated in a batch of seedlings that Porter grew in the late fifties or early sixties. He placed little value on it because the flowers were sterile. But an American nurseryman, touring with the Great Plains Horticultural Association one spring, leapt at the chance to propagate it. Here was the perfect tree for those gardeners who hated "mess"—no crabapples would fall on deck or lawn. He asked permission to introduce the tree into the U.S. nursery trade and wondered if Porter would suggest a name. Porter suggested 'Summer Snow.' As Allan Daku noted, "Our American neighbors do not find the same fascination with snow that we do and particularly not in summer. They countered with 'Spring Snow' and it came into being." (*Friends' Newsletter*, 2004)

Porter received a royalty of ten cents per tree for this introduction. A cheque arrived on June 28, 1968 for $453.50.

Roses

Porter introduced three roses: 'Shellbrook,' a form of *Rosa acicularis* found near Shellbrook, with deep red blooms followed by long, bottle-shaped hips; 'Parkside,' with single deep pink flowers found near Parkside, 2 m in height; and the 'Double Altai' shrub rose. None are now available.

Lilies

Although Porter had offered both the single and double forms of the tiger lily in his 1940 and 1942 listings, the real impetus toward lily hybridizing came in the fall of 1948, when his friend and fellow nurseryman Percy Wright, then living at Moose Range near Carrot River, gave him some lily seed as well as a box of lily bulbs.

Porter started the seed and planted the lilies. They flourished in Honeywood's light sandy soil, and he found the large flowers easy to hybridize. He was hooked. He also used Skinner and Preston lilies in his early breeding. By 1952, Honeywood was offering a collection of its own unnamed lilies. A year later, there were six named lilies in the catalogue as well as a European species.

Christmas cards from around the world arrived at Honeywood. Courtesy Honeywood Heritage Nursery.

The first mimeographed lily price list in the fall of 1954, and the first printed list in the fall of 1956 included Porter's first four lilies, hybrids of Patterson and other Canadian hybridizers, as well as six species lilies.

He wrote articles to popularize the growing of lilies and urged readers to join a lily society. "...Of the eighty or so known species of lilies, at least twenty... are hardy here and many of these make good garden subjects... These lilies are as easy to grow as potatoes... Some lilies will wander underground for some distance, sometimes a foot or more, before emerging in the spring. Thus one should be careful with the hoe." (Porter, 1971 *Prairie Garden)*

Porter felt strongly that "no lily should be named unless it is really unique in some respect.... It should also be vigorous, disease-resistant and easily propagated." (Porter, "Progress in Breeding Hardy Lilies") Hardiness, of course, was essential. "In our breeding work at Parkside we are continually looking for the best forms, the clearest colours, and earlier and later blooming types than we now have. Another thing we're beginning to look for is fragrance... Our prairie lily, *L. philadelphicum andinum*, has a delicate and delightful odor—if one takes the trouble to sniff." (A.J. Porter, "Progress in Breeding Hardy Lilies") By the 1980s he was also selecting dwarf lilies that would do well in pots or for forcing.

Networking

Although geographically isolated, Bert Porter carried on an extensive correspondence both in North America and internationally. As he often remarked, the mails were a lot more reliable then. He exchanged letters, bulbs, and seeds with other Canadian breeders such as Cecil Patterson, Frank Skinner, Isabella Preston, Robert Simonet, Fred Fellner, and Percy Wright. Theirs was a fraternity in which advice and breeding material went freely through the mail.

In 1961, Porter wrote to Jan de Graaf of the Oregon Bulb Farms in Gresham, Oregon, in response to an article de Graaf had written on hardy lilies in the *Prairie Garden*. He chided de Graaf for failing to mention "that there can be quite a difference in hardiness within a species, depending on whether the original stock came from the southern extreme or the northern extreme of its range. In mountain species it could be that those growing up toward the snowline would carry greater hardiness to freezing than those found in the valley."

Porter also expounded the advantages of the deep planting of lilies: "We have a deep black sandy loam that warms up quickly in the spring so have found it an advantage, especially with the martagon hybrids, to plant deeply so as to retard their emergence in the spring. Some lilies give a better crop of stem bulblets for the simple reason that there is more stem underground to bear the bulblets, but it does not seem to detract from the main bulb."

He concluded with an invitation to "come on up and have a look at the lilies here. We are eighty miles north of Saskatoon, forty odd miles west of Prince Albert, about three hundred miles

Lily Introductions

The following Porter introductions are in the Memorial Garden at Honeywood and should be available for sale within a few years: 'Cheerful Charlie,' 'Raspberry Rhime,' 'White Picture,' 'Sky Dancer,' 'Saskatoon,' 'Golden Dauricum,' 'Red Carpet,' ('Edith Cecilia' x 'Rosabelle #5'), ('Edith Cecilia' x 'Rosabelle #2'), 'Sunshinebright,' and 'Princess Royal.' Porter also introduced 'Dark Gleam,' 'Firebright,' and 'Pink Chartreuse,' which at the time of writing appear to be lost. The following remain available from Honeywood, the Canadian Prairie Lily Society (www.prairielily.ca), and other lily nurseries.

'Earlibird' (1956)—out-facing, clear vivid apricot, fine dark spots near centre; early; 80 cm; resistant to botrytis; durable garden lily.

'Rusty' (1956) (seed of Isabella Preston hybrids)—up-facing, rusty orange, spots, almost flat blooms; early; 60 cm.

'Rosabelle' (1956) (*L tigrinum* hybrid)—up-facing, star-shaped, deep old rose, spotted maroon; July; 100 cm.

'Golden Jubilee' (1956)—up-facing, rich golden yellow, good substance; June; 100–120 cm.

'Delicious' (1961) (seed of Isabella Preston hybrid)—up-facing, gold-orange, heavily spotted, somewhat reflexed petals; July; 100 cm; slight fragrance; NALS Award of Commendation 1967.

'Redland' (1962) (Percy Wright seed pod x 'Gold Urn'; syn. 'Warrior')—up-facing, large, deep velvet red, no spots; July; 120 cm; 1971 WCSH Merit Award.

'Pink Champagne' (1963) (*L. tigrinum* hybrid)—out- and down-facing, yellow with pink overlay; late July/mid-August; 120 cm.

'Orange Light' (1964) (Percy Wright seed pod x 'Gold Urn')—up-facing, cup-shaped blooms, fluorescent orange, nearly spotless; June; 100 cm; WCSH Merit Award 1971.

'Flaming Giant' (1975)—up-facing, large, cup-shaped flared blooms, vivid red-orange, spots; July; 120–150 cm; good substance.

'Geisha Girl' (1975)—out- and down-facing, lemon-yellow with spots; July; 80 cm.

'Scarlett' (1975)—down-facing, spotless orange-red, reflexed petals; July; 100 cm.

'Jolly Miller' (1979)—up-facing, buttery yellow with few if any spots; good substance; late June–early July; 50–60 cm.

'Cinnamon Toast' (1984)—up-facing, clear white with cinnamon spots; July; 90 cm.

'Golden Age' (1984)—up-facing, golden yellow; dark bronzy stems; July–August; 75–90 cm; anniversary namesake.

'Isis' (1986) (syn. 'Cover Girl')—up-facing, soft ivory yellow with a tint of green toward centre, scattered spots, flowers open flat; July; 90 cm.

'Gale's Favourite' (1986)—up-facing, rich bronze-red, black stems; July; 90 cm.

'Happy Thoughts' (1986)—up-facing, pure white, almost spotless, hint of green in throat, ivory stripe along mid-rib of each petal, candelabra type inflorescence; July; 75–100 cm.

'Susan Eileen' (1986)—up-facing, bowl-shaped flower, apricot washed with delicate pink; July; 90 cm.

'Glistening Star' (1995)—down-facing, star-shaped, glistening orange, reflexed petals; June; 90–100 cm.

north of the U.S. boundary."

Honeywood would eventually sell lily bulbs to Argentina, Czechoslovakia, Germany, South Africa, New Zealand, Australia, Denmark, Britain, Japan, Sweden, and the Soviet Union, as well as Canada and the United States.

By the early 1960s many international orders were arriving, including some from behind the Iron Curtain. These were often followed by Christmas cards. As Allan Daku wrote, "These folks cannot send currencies to the west, so strike some rather interesting exchanges of merchandise for lily bulbs." (Daku, "Honeywood Nursery") One such customer wrote, "I can send you for it some interesting Russian souvenirs." Porter once received an absolutely gorgeous set of eight Matryoshka dolls, nestled inside of each other. On other occasions, a red velveteen photo album and a book on Russian porcelain (with an English translation) arrived.

One of these correspondents was E. B. Kieslyk from Alma-Ata, USSR. "Your lilies blossomed very well. All the kinds are really wonderful; and I can hardly say what of them are the best. I was alarmed with the color of the lily 'Black Butterfly.' Other lilies have unusual colors too."

Vaclav Jost, who worked at a State Breeding Station in Czechoslovakia, asked Porter to send a catalogue "in which I can learn your sortiment of lilies and perhaps order some interesant kinds."

From Flubert Majchrzyk of Poland in 1979: "Please, can you send *gratis* a piece of bulbs or seeds of *Lilium amabile, L. concolor, L. davidii*, and *L. philadelphicum*. I am an amateur gardener and like lilies. Yours sincerely."

The Lily Robin

Through the early 1970s until at least the late 1980s, Porter was

Honours

Bert Porter received many honours through his lifetime, especially considering his lack of formal training and the geographic isolation in which he worked. He was recognized not only in the field of horticulture but by his community.

George Krahn

Dr. Porter after receiving his honorary doctor of laws degree, University of Saskatchewan, 1983. Courtesy Rich Krahn.

Saskatchewan Horticultural Association, honorary life member (1955), Certificate of Merit (1968), president, 1969, 1970)
Manitoba Horticultural Society, Stevenson Gold Medal (1963)
Massachusetts Horticultural Society Gold Medal (1969)
North American Lily Society, E. H. Wilson Memorial Award (1971) "for his noteworthy contribution to the development of lilies. It is the highest award given by this Society."
Saskatchewan Nursery Trades Association, honorary life member (1977), president (1965, 1971), secretary, founding member
German Garden Festival, Stuttgart, Silver Medal (1977)
Walter Willoughby Horticultural Society (charter member, president, treasurer), honorary life member (1977)
D. R. Robinson Award, Saskatchewan Horticultural Association (1975)
Canadian Nursery Trades Association, honorary life member (1978)
Saskatchewan Institute of Agrology, honorary life member (1979)
University of Saskatchewan, Honorary Doctor of Laws (1983)
Rural Municipality of Leask, Appreciation Award (1983)
Parkside community, Silver Platter (1983)
Western Canadian Society for Horticulture, honorary life member (1983), Merit Awards
Canadian Association for Horticultural Science, Special Recognition Award (1986)
Canadian Prairie Lily Society, founding member, honorary life member (1987)
Saskatchewan Agricultural Hall of Fame, inductee (1987)

Community Service

Bert Porter was president of the Shellbrook Rural Telephone Company for twenty-five years; it was the last local service in the province to be taken over by SaskTel. He was a founder of the Shellbrook Credit Union and a board member for eighteen years. He was trustee and secretary of the Honeywood School District and was president of the Parkland School Trustee Association.

Porter judged in many local, provincial, and national horticultural and lily shows, finally retiring from judging in 1989. He was a participant in the first Provincial Fruit Show and many thereafter and was well known for his prize-winning entries. As well, he taught workshops in landscaping, gardening, and judging in various communities and worked with local 4-H Garden Clubs. In 1988, he was a founding member of the Saskatchewan Fruit Growers Association.

part of an informal "Round Robin" of fellow lily hybridizers, members of the North American Lily Society living in Canada and the United States. They exchanged ideas, information, seed, bulbs, slides, and photos of their work.

On October 11, 1971, he wrote: "Dear Robins: My face is very red. My wife just pulled out the bottom bureau drawer and found at the back of it the lily Robin envelope with all the letters, pictures and slides. This had evidently arrived in August when I was away and had somehow slid down behind the bureau on which my wife had placed it awaiting my return... Lucky she took out that bottom drawer or we wouldn't have known until house cleaning time next spring what had become of the Robin....My most sincere apologies ... I think the highlight of the summer was the visit from Mrs. Biddle [who personally presented Porter with the E. H. Wilson Award] and Miss Rosseter. They were very enthusiastic about my new seedlings... I sold most of my lily blooms when they were in full bloom. Visitors took them along right away. It sure saved a lot of packing this fall. One lady was visiting here from Michigan. She took back a whole trunk full, all in full bloom and reported the plants kept right on blooming and hardly showed that they had been moved."

By the 1970s, between the last week of September and the first week of October, he would mail between five thousand and ten thousand lily bulbs. (Russell)

Honeywood Heritage Nursery

In the fall of 1999, a group of sixteen long-term community members purchased the nursery and the quarter section of land on which it stood to save Bert Porter's living legacy and rebuild the nursery as a continuing community heritage operation. Their goals were to reclaim the nursery grounds, restore the Porter home, and locate and acquire plants introduced by Dr. Porter. One of the top priorities was to turn the old log home into an interpretive centre and tea room. Their mission is to preserve, restore and operate Honeywood Nursery as a Parkside Community heritage site. In the spring of 2001, it was designated a municipal heritage site in Saskatchewan.

Under the energetic and capable direction of manager Judy Harley, and countless hours of volunteer work from members of the community as well as generous corporate support, the site is being returned to its former glory, and many of Bert Porter's introductions are being returned to the nursery by generous donors nationwide. (Contact: P.O. Box 48, Parkside, SK, S0J 2A0, www.honeywood-lilies.ca).

On December 3, 1972, he reported in the "Robin": "We finally have all the parcels shipped out and the remaining bulbs put to bed for the winter. We again had an increase in orders this season, so it really kept things humming here to get them all packed and away. Our weather through September and October was miserable. Two days in September the temperature stayed below the freezing mark. We had snow in September and again in early October. Seemed we would get one nice day and then it either rained or snowed or froze or the wind blew a gale all day, making it very difficult digging bulbs."

Articles

Bert Porter wrote numerous articles for the *Prairie Garden* annual from Winnipeg, as well as more scientific ones for his professional colleagues. He reported regularly in the Proceedings of the Western Canadian Society for Horticulture on the progress of the fruit and lilies being grown at Honeywood, his own introductions as well as those of others.

He also gave advice about removing lily pollen in the December 1972 "Robin": "Tell your lady friends on no account to rub it when they get it on their clothes. This just smears it in. Let it dry thoroughly and then just brush it off. There is no need to ruin dresses if this is done."

On June 12, 1989, he wrote about having thought he had sent on the "Robin" but had instead misplaced it. "I most sincerely apologize. I think, that I am getting so forgetful the Director had better eliminate me from the list. I will certainly miss receiving the letters. But it is inexcusable for me to have kept the Robin so long." His fellow Robins would not think of such an "elimination," and he continued on the list.

For more than three decades, Porter regularly sent money orders to London to register his lilies. In a reply dated July 30, 1974, J. W. O. Platt, Secretary, Lily Committee, Royal Horticultural Society, wrote, "Thank you for the description of your lily clone 'Red Carpet' and the money order for 63 p. I have registered the name which will appear in the Supplement to the International Lily Register to be published in "Lilies 1974" in the autumn."

Other Breeders' Introductions

As Bill Cumming wrote of Porter in the 1978 North American Lily Society *Yearbook*, "He is not only a breeder and introducer par excellence, but has made many of the products of other lily breeders available to the general public. These include the Patterson hybrids, Barber lilies from Elfros, Saskatchewan, and several lilies bred by the very successful and well known breeder of horticultural plants, Mr. Robert Simonet of Edmonton, Alberta."

Porter had obtained the Barber lilies from the Horticultural Society in Elfros where they had been originally raised by Mr. Barber. In the 1967 WCSH Proceedings, Porter wrote, "They are said to be derived from Mr. Patterson's hybrids but the blooms are of better substance and have wider tepals, many of which are entirely free of spots, giving clearer colors". Porter eventually introduced 'Barber #17,' 'Lady Dawn,' 'Suntan,' 'Make Believe,' 'Honeybunch,' and 'Girls Only.'

From Robert Simonet came 'Black Butterfly' and 'Embarrassment,' both of which won NALS Awards of Commendation, and 'Summer Night.' And from Percy Wright came 'Red Knight,' 'Flame,' and 'Towering Turk.' Last came 'Ed Brooman' and 'Sungod,' hybridized by Ed Brooman.

Porter's own lilies were also widely used by other breeders. It has been estimated there are thousands of lilies in Canada and abroad that owe their origins to Honeywood Nursery and Bert Porter.

Sources

Bert Porter obituary, undated, no source.

Brown, Jane, "A. J. Porter: A Lily Legend," *Rural Roots*, 7 Oct. 1993.

Cumming, W. A., North American Lily Society *Yearbook*, 1978.

Cypher, Dick, "A Visit with Winnie," Friends' Newsletter, Honeywood Heritage Nursery, 22 Nov. 2005.

Daku, Allan, "Honeywood Lilies," undated.

———, "Honeywood Nursery," *Reflections of the Past*, Parkside Historical Society, 1991.

———, "The Old Root Cellar," *Friends' Newsletter*, Honeywood Heritage Nursery, 22 Nov. 2005.

———, "The 'Spring Snow' Crab," *Friends' Newsletter*, Honeywood Heritage

Nursery, April, May, and June 2004.
———, e-mail to author, January 2006.
Doak, George O., letter to Bert Porter, 14 July 1971.
Harley, Judy, "Porter Lilies and Honeywood Introductions," undated.
———, e-mail to author, January 2006.
Honeywood catalogues: 1940, 1963, 1969, 1972, 1984, 1986.
North American Lily Society Hybridizers' "Round Robin" #3: 11 Oct. 1971, 11 June 1972, 3 Dec. 1972, 12 June 1989.
Jost, Vaclav, letter to Bert Porter, undated.
Kadota, Diane, "Porter's Lilies," *The Western Producer,* 22 Sept. 1983.
Kieslyk, E. B., letters to Bert Porter, 4 Sept. 1970, 10 Jan. 1972, 13 Dec. 1972.
Krahn, George, Saskatchewan Nursery Trades Association, Landscape Saskatchewan, Historical Review, 1957–1982, 1986.
Majchrsyk, Flubert, letter to Bert Porter, 1979.
Moe, Dave, e-mail, Jan. 2006.
Parkside Historical Society , "Reflections of the Past," 1991.
Platt, J. W. O., letter to Bert Porter, 30 July 1974.
Porter, A. J. (Bert), "Instructions for Pickers," undated.
———, letter to Jean Erickson, 16 July 1954.
———, "Progress in Breeding Hardy Lilies," undated.
———, letter to Jan de Graaf, 11 March 1961.
Porter, Thomas, "A Honey of a Wood," *Saskatchewan Naturally* 3, 1 (Spring 2001).
Porter, Winifred R., "Porter, A. J. (Bert) and Winnie," Parkside Historical Society, *Reflections of the Past,* 1991.
Prairie Garden, 1963, 1964, 1969, 1971, 1985.
Russell, Nancy, "Lily Business is Blooming at Parkside Nursery," Saskatoon *Star-Phoenix,* 21 June 1977.
"Saskatchewanians Honoured," *The Western Producer,* 12 May 1983.
Sjulin, Larry, Inter-State Nurseries, letter to Bert Porter, 28 June 1968.
Shellbrook *Chronicle,* Jubilee Edition, May 1955.
Stupple, Gwen, "Consider the Lilies… A Thumbnail Sketch of Bert Porter," *Landscape,* 1977.
Toop, Betty, "Memories of Porter's Place in the Forties," *Friends' Newsletter,* Honeywood Heritage Nursery, 14 July 2005.
Valleau, John, e-mail to author, 2003.
Western Canadian Society for Horticulture, Proceedings: 1945, 1954, 1967, 1970, 1972, 1984.
Williams, Sara, "A. J. (Bert) Porter—Plant Breeder for the Prairies," *Prairie Garden,* 1985.
———, "Fruit Introductions of A. J. Porter (1933–1983)," March 1983, unpublished.

Les Kerr (1902–1983)

An avid hunter, fisher, and bird watcher, Les Kerr was also an ardent conservationist and collector: of coyote pups, bear cubs, baby raccoons, and for a while, a talking magpie—animals that would one day form part of the nucleus that would become the Forestry Farm Zoo in Saskatoon.

As a horticulturist, his legacy was varied and lasting: plums, cherry-plums, apricots, the 'Kerr' crabapple, flowering crabapples, caragana, and elders. Less well known was his continued support of the horticultural industry. He provided the impetus to organize the Saskatchewan Nursery Trades Association and assisted them technically for decades.

And two young boys, evacuees from the bombing of Britain, will always remember the Kerrs as their surrogate Canadian parents during the war years.

William Leslie Kerr was born near Renfrew, Ontario in 1902. His father was a nurseryman, and Les developed a love of plants and outdoor life at an early age. He received his Bachelor of Science degree in Agriculture from the Ontario Agricultural College in Guelph, Ontario in 1925 and a Master of Science from the University of Maryland in 1928. He remained in Maryland for four more years, from 1928 to 1932, working for the Federal–State Fruit and Vegetable Inspection Service.

Returning to Canada in 1932, he took the position of officer in charge of fruit breeding for the Canada Department of Agriculture at the Rosthern Experimental Farm in Saskatchewan. Later that same year, he was transferred to the Morden Experimental Station

"I would like to leave this thought. It will be our soils, forests and waters which will pay back our natural debts. No manipulation by our good friends the bankers, bonds-men or lawyers can do this. Our trees hold the key to much pleasure, comfort and prosperity."

Les Kerr, c. 1920s. Photo PH951571 courtesy of the Saskatoon Public Library Local History Room.

in Manitoba. He remained at Morden from 1932 until 1942, first as research officer in charge of fruit breeding and later as assistant superintendent.

In 1942, Les Kerr became the superintendent of the Sutherland Forest Nursery Station, a few miles northeast of Saskatoon. It would later become the Prairie Farm Rehabilitation Administration (PFRA) Tree Nursery. He retired as a federal employee in 1966 when the PFRA Sutherland Tree Nursery was officially closed and the west 120 acres donated to the City of Saskatoon as a park. But he continued to work as supervisor of the Forestry Farm Park for the City until 1970. His older brother Harold, a returned army man, lived in the log cabin just west of the main entrance, selling ice cream and other treats during this period.

Les Kerr supervised the production and distribution of millions of trees to farmers across the prairies from the Sutherland Forest Nursery Station. Between 1914 and 1963, 147 million trees were distributed from Sutherland. According to the Dominion Forest Nursery Stations Progress Report, 1947–1952, "An annual average

Active in the Community

During his time at Sutherland, Les Kerr was active in the Saskatoon Horticultural Society and as a Saskatchewan Horticultural Association director and show judge, and promoter of the Provincial Fruit Show.

He also served his community in other capacities. He was a member and director of the Canadian National Institute for the Blind (where he did a lot of volunteer pruning on his own time), the Rotary Club, Kiwanis, the Canadian Club, and the Saskatoon Fair Board.

He was chairman of the provincial game farm management committee, a charter member (1948) of the Saskatoon Fish and Game League (which later became the Saskatchewan Wildlife Federation), and chairman of the Beaver Creek Pheasant Farm Committee, which oversaw the raising and release of pheasants from 1944 to 1966.

Les Kerr died on May 28, 1983 at the age of eighty-one following a series of strokes. He was survived by his wife, Blanche. After a funeral service at Grace Westminster United Church, his ashes were spread by airplane above Emma Lake near his beloved cabin.

Les Kerr Park in the Forest Grove area of Saskatoon was named in his honour.

Blanche Luke and Les Kerr were married in Rosthern in 1936. From Rosthern Historical Society, Old & New Furrows, The Story of Rosthern, p. 476. Photo LH971.242 R839 courtesy of the Saskatoon Public Library Local History Room.

Blanche Luke Kerr

Les Kerr met Blanche Luke, a teacher, while he was working in Rosthern. Born and educated there, she was one of four daughters of James (Jim) and May (née Palmer) Luke. These young women loved horses and grew up in a sociable household where friends were always welcomed to gatherings around the piano and there was always room at the table for an extra guest.

Blanche married Les Kerr in 1936 and they moved to Morden, where he was then stationed. She had always admired the brick superintendent's house at the forestry station in Sutherland but had never dreamed she'd ever live in it. When Kerr became superintendent of the Forestry Station in Sutherland, she settled happily into the large brick house in its parkland setting.

Superintendent's residence, Forest Nursery Station, c. 1916. Courtesy Friends of the Forestry Farm House.

of 19 refrigerator cars were required to accommodate trees shipped" in the late forties and early fifties.

He tested hundreds of new species and cultivars for their suitability as shelterbelt trees. As well, he carried out extensive research on insect and disease problems, cultural practices such as irrigation and fertilizing, rodent and pest control, overwintering storage methods and facilities, and both seed and vegetative propagation methods.

Extension

Les Kerr was keenly aware of the value of extending the research carried out at the station into the prairie community. He wrote articles on various aspects of shelterbelt planting and design, improving the home landscape, propagation, and other aspects of horticulture. He recognized the value of shelterbelts for water and wildlife conservation, erosion and flood control. Many of the more technical presentations originally made for his peers from research institutions or the nursery trade were quickly made available to the general public as newspaper or magazine articles. In 1947, he was a co-author of a booklet on the budding and grafting of fruit trees published by the federal Department of Agriculture.

In 1944, World War II had not yet ended. The national conscience was still focussed on the war effort, and many household breadwinners were in uniform. Yet Kerr was able to look forward to a time when forests would need legislative protection from industrial exploitation, long before the term "renewable resource" was coined or multinational companies were the norm. In a 1944 speech, Kerr emphasized the importance of natural forests and trees: "A sound forestry policy should be initiated to make the maximum use of it by a balance of controlled commercial exploitation, reforestation and conservation. We should not have lumber ghost towns, soil erosion, and destructive floods as experienced in many parts of the world if we act in time."

And he was able to make the transition from a global perspective to an urban or rural landscape. "The importance of trees in making our homes more pleasant and inviting should not be overlooked. Who wants to live on a treeless farm or in a

The Forestry Farm Park

Kerr spent a lot of his weekends and holidays at his cottage at Neis Beach on Emma Lake, where he fished and observed birds. After his retirement, he used his cottage property to continue his plant breeding.

He was also an avid hunter. It is no coincidence that what began as a tree nursery ended up as a municipal zoo. Many an orphaned animal found near Emma Lake eventually found a home at the Sutherland Station.

He befriended a porcupine that later gave birth. A tame coyote eventually ended up with a film career in Hollywood. He crossed other pet coyotes with a Labrador dog. His interest in genetics led him to breed a wolf-dog cross with a dingo. He brought home motherless bear cubs from Emma Lake and fostered fawns, baby raccoons, and even had a talking magpie at one time. These formed the nucleus of an informal collection of native animals that together with the animals from the Golden Gate Animal Farm would officially open as a zoo in 1972. Although his federal superiors were not always happy with the animals at the station, the Saskatoon *Star-Phoenix* pointed out that "It was a logical combination for a horticulturist who also happens to be an ardent conservationist and animal lover." (4 June 1966)

Les Kerr at Sutherland Tree Nursery with doe and fawns. Photo LH9460 courtesy of the Saskatoon Public Library Local History Room.

Blanche and Les Kerr with their "adopted" son, Peter Wilde, Sutherland, c. 1943. Photo PH951584 courtesy of the Saskatoon Public Library Local History Room.

treeless town? No matter how nice the house may be it is not a home until there are trees around it. The relationship of trees to the abundance and variety of birds and animals is of major importance."

He continued the work of his predecessors in making the Sutherland Forest Station itself a living model of what a rural landscape could be. As Don Dabbs wrote in his 1982 convocation speech when Kerr received his honorary degree, "Uncounted hours of his own free time were devoted to the development of the beautifully landscaped grounds of the Forestry Nursery Station and to the development of an excellent animal park within these grounds." An article in the Saskatoon *Star-Phoenix* in June 1966 concurred: "Under Mr. Kerr's green thumb, the Sutherland Station became a Mecca for farmers, gardeners, horticulturists, nursery operators and sightseers from both sides of the border."

Wildlife Conservation

While never neglecting his official mandate, Kerr increasingly emphasized the value of shelterbelts and trees for wildlife conservation. He noted that birds added interest to our landscapes while playing an important role in insect control.

As chairman of the WCSH Rodent Control Committee for many years, he was well aware of what is now called the "food chain." He realized the value of coyotes and foxes in the control of mice, rabbits, and pocket gophers.

He was not only observant but realized that much harm could come from the arbitrary elimination of predators. Nor was he a man without a sense of humour. In 1951 he wrote in the WCSH Proceedings: "There is some indication that rodents such as field mice, bush rabbits and jack rabbits are on the increase in many districts... There are large areas now where their predators have been eliminated. We may some day regret such effective control measures and prefer to have the odd chicken thief around rather than be overrun by rodent pests."

In his annual reports and in articles in various publications, he listed species and varieties of trees and shrubs attractive to birds. He recommended the development of public wildlife plantings

Les Kerr with 'Sutherland' caragana measuring depth of snow. Courtesy Rich Krahn.

to provide habitat while adding to the attractiveness of the rural landscape. He felt these were especially needed on wasteland adjacent to towns and cities and noted that both individuals and organizations could become involved in their establishment.

Special mention was made of trees and shrubs that produce edible fruits for wildlife in fall and winter such as American mountain ash, bur oak, chokecherry, European red elder, hawthorn, highbush cranberry, Nanking cherry, pincherry, roses, rosybloom crabapples, sandcherry, and wild plum.

Work with Nursery Trades

Growing up on his father's nursery in Renfrew, Ontario, "It seemed only natural for this man to express a personal interest in the welfare of the nurserymen with whom he so freely exchanged views and experiences." As George Krahn recalled in 1999, "He stopped by the nursery frequently, always with good advice... Les was a man of few words until he got to know you. So initially,

Les and Blanche Kerr with Peter and Douglas Wilde, Morden, c. 1941. Photo PH951581 courtesy of the Saskatoon Public Library Local History Room.

Guest Children

In 1941, the Kerrs read an ad in a local Morden newspaper asking for surrogate parents to look after "guest children," evacuees from British cities vulnerable to enemy bombing. As they had no family of their own, they applied for a child. Peter and Douglas Wilde, brothers seven and twelve years of age respectively, arrived later in the same year, each "clutching a satchel of clothes and a few personal necessities." (Melberg) They were two of thousands of such children evacuated to Canada.

Peter, the younger brother, left with the Kerrs for Saskatoon when Les Kerr became superintendent of the Forestry Farm. He recalls that they needed special permission to move him. The older of the brothers, Douglas, went to another family in Morden for the duration. Both boys kept in touch with their Canadian "parents" and visited with them many years later.

Returning to Saskatoon in 1981 (thirty-five years after he'd left) for a reunion with the Kerrs, Peter Wilde's first impression was that his "mother" was "a lot shorter than I remember."

"We did go down to the old house in Sutherland, saw the places where I used to ride the neighbor's pony, where I used to go hunting and we all remembered where I got all tangled up in the barbed wire fence and ended up with stitches." (Melberg)

Douglas Wilde, in a memoir written in 2003, recalled the Morden of 1941: the timber plank sidewalks, the outdoor privy, the pump in the kitchen that drew water from a well under the house. "On bath days I guess about once a week a zinc bath was brought in from the rear porch and placed on the kitchen floor. It was then filled with water that had been heated on the stove, then it was Pete my brother at one end and me at the other—no worrying about taps digging in your back!" Drinking water was fetched from a pump at the end of the street. "Placing the wood pile between the privy and the back door became relevant to my own experience, you brought logs back to the house on your return journey."

Les Kerr later confided to a friend that had it been possible, they would have permanently adopted Peter.

When Peter returned to visit in 1981, Blanche said, "Les had always told me to go to England to see Peter, but it was always just too far away. Les had a cabin at Emma Lake and doesn't like to go too far from there." (Melberg) After Les Kerr's death, she did make the trip to England and was warmly welcomed by both of her now grown "sons."

Honours and Volunteer Service

Saskatchewan Fish & Game League, honorary life member, 1953

Kiwanis Club, Distinguished Service Award, 1961

Canadian Society for Horticultural Science, honorary life member, 1966

Western Canadian Society for Horticulture, past president (1954), charter member, honorary life member, 1969

Stevenson Memorial Gold Medal, 1970, for "conspicuous achievement in the field of practical horticulture" and "for his imaginative and practical pursuit of the breeding and propagation of prairie-hardy fruits, trees, and ornamental shrubs." Kerr received the last medal actually made of gold.

Saskatchewan Nursery Trades Association, honorary life member, 1977

Saskatchewan Wildlife Federation, honorary life membership, 1977

David R. Robinson Award, Saskatchewan Horticultural Association, 1978

University of Saskatchewan, Honorary Doctor of Laws "in recognition of his contributions to western Canadian horticulture and the improvement of our quality of life," Spring 1982.

Saskatchewan Horticultural Association, honorary life member, director

Great Plains Section of the American Society for Horticulture Science, president

Introductions

Prunus

Kerr was especially interested in the genus *Prunus*, which includes plum, cherry-plum, mayday, chokecherry, flowering almond, Nanking cherry, Mongolian cherry, and apricot. As early as 1937, he wrote a scientific paper on *Prunus* hybridization.

Mayday-chokecherry

By 1964, Kerr had selected non-suckering rootstock of mayday *(Prunus padus)* onto which he top-worked chokecherry for shelterbelts.

Of his mayday-chokecherry selections, 'Maxi' and 'Sharon' chokecherry and 'May Queen' mayday, none are readily available today. Yet all of these had been considered worthy enough to enter in the 1980 Western Canadian Society for Horticulture (WCSH) merit trials.

Cherry-plums

Kerr introduced two cherry-plums, both 'Sapa' seedlings. 'Manor' is readily available today, while 'Dura' is more difficult to locate.

'Dura' (1942, Morden)—fruit 3.5 cm; skin dark green mottled in purple; flesh red-purple; ripens late August; good for dessert and canning; low spreading bush.

'Manor' (1945, Morden)—fruit 2.5–3 cm; skin dark reddish-purple (almost black) with a light waxy bloom; flesh dark red, nearly freestone, sweet and juicy; ripens mid-August; good for fresh eating and jam; tree 3 x 2 m, upright and bushy.

Plums

Of the three plums introduced by Kerr, two—'Bounty' and 'Norther'—are still sold. 'Kerr's Gold' is no longer available.

'Bounty' (1939)—fruit 3.8 cm; skin dark red with a blue bloom, tough, astringent; flesh orange-yellow, tender, juicy, and sweet; ripens mid-August; good for preserves; tree upright and spreading.

'Norther' (1940) ('Assiniboine' seedling)—fruit 3.5 cm; skin bright red; flesh yellow, sweet, and juicy; ripens early August; good for fresh eating.

Western sandcherry–Nanking cherry hybrids

Kerr also developed hybrids of western sandcherry *(Prunus besseyi)* and Nanking cherry *(P. tomentosa)*. As Percy Wright wrote in a 1972 article in *Horticulture*, "The project is typical of the numerous imaginative plant-breeding programs undertaken by Mr. Kerr. *P. besseyi* is drought-tolerant, hardy, and will set fruit even in highly unfavorable seasons, but it has a prostrate form and poor fruit. *P. tomentosa* has more valuable fruit but blooms early and is often caught by frost. The hybrids so far have either resembled one parent or the other rather than being intermediate." Ultimately, Kerr developed an improved Nanking cherry with larger fruit in either white, black, red, or maroon, depending on the seedling.

Les Kerr measuring growth of young tree c. late 1920s. Photo PH951572 courtesy of the Saskatoon Public Library Local History Room

Apricots

Like his work with plums and cherry-plums, Kerr's interest in apricots began while still at Morden. In 1935 he reported that "early maturity and hardiness at this Station warrants the inclusion of apricots to the promising fruits for the prairies." In 1936, he added: "It should be noted here that apricots offer much promise. No tree fruits withstood the hot dry weather and spider [mite] infestation so well. Great variation and extreme winter hardiness is evident among the seedlings. The compatibility and inter-crossing with many other stone fruits may be a valuable trait."

Both of Kerr's apricot introductions, 'Scout' and 'Wescot,' are still recommended and available.

'Scout' (1937) ('McClure' and *P. mandschurica)*—fruit 3.8 cm; skin bronze gold blushed with red; flesh sweet; ripens early August; dries well and good for canning, jam, and fresh eating; tree 4.5 m.

'Westcot' (1982) *(Prunus armeniaca)*, (Morden); fruit 4.4 x 3.5 cm; skin yellow orange with light reddish blush; flesh orange, fine textured, juicy; mild sweet apricot flavour; free-stone; greater bud hardiness and quality than 'Scout.'

Just prior to his death in 1983, Kerr donated two thousand dollars to the Department of Horticulture Science, University of Saskatchewan, to support graduate student research on improving *Prunus* species for northern areas. That gift has recently begun to bear fruit in the development of hardy, long-stemmed dwarf cherries for fresh eating and processing.

Crabapples

'Kerr' (1952) ('Haralson' x 'Dolgo') (Morden)—fruit 4.5 cm; skin creamy yellow with dark reddish-purple blush; flesh yellow to creamy white, juicy, firm and crisp and sweetly tart; ripens mid-September; good for fresh eating, cooking and canning and keeps up to twenty-seven weeks; tree is broadly upright with a round head and very sturdy. 'Kerr' is still popular and available a half a century after its introduction.

Rosybloom flowering crab-apples

Kerr began his breeding work with rosybloom crabapples while at Morden. By the early 1940s, seed of some of his better selections from Morden had been planted at Sutherland. His breeding objectives were improved tree form, fruit, bloom, foliage, and disease-resistance. George Krahn recalled that Kerr's real pride was the one kilometre of rosybloom crabapple trees from which he would eventually make named selections. Krahn helped with the budding of many of these.

Among the selections eventually made were 'Almey' (1945), 'Pink Cascade' (1964), 'Sutherland' (1955), 'Royalty', 'Dainty' (1963), 'Fuchsia Girl,' and 'Pink Spires.' Of these, 'Royalty,' 'Fuchsia Girl,' and 'Pink Spires' are still available.

In a 1965 *Prairie Garden* article, Kerr noted that 'Sutherland' "transmitted its dark foliage and bloom colour to its offspring more liberally than any other selection. In stature they range from extreme dwarfs to tall vigorous trees. In form, the range varies from weeping to pyramidal. Foliage colour varies from greenish to very dark reddish purple. Many have distinctive two-toned leaves. In other words, it is the plant breeder's dream for variety and distinctiveness." As a result, he used it for later selections, even after it had been long superseded.

'Royalty' (1958) ('Sutherland' x 'Rudolph')—very glossy dark reddish-purple foliage; large single reddish-pink flowers with wide petals; persistent purple oval fruit, on long stems; tall upright tree to 6 m. Selected as one of Canada's centennial trees, 1967; WCSH Merit Award, 1968.

In 1999, George Krahn recalled the first time he saw 'Royalty' at the Sutherland Station: "Finally, we came to the last selection. I was astounded. Here was a tree with deep reddish-purple leaves, lustrous in color. It had, at that time, the darkest red foliage of any hardy crabapple in Saskatchewan."

Widely grown beyond the prairies, its selection as a Canadian centennial tree resulted in a very heavy demand for propagation material. In 1966, 40,000 buds of 'Royalty' were shipped to government institutions and commercial nurseries. But like its parent 'Sutherland,' it proved susceptible to fireblight.

'Fuchsia Girl' ('Royalty' seedling, 1998)—pink-red flowers; deep purple foliage; dark red fruit; less susceptible to fireblight; upright form (5 x 5 m).

'Pink Spire'—single lavender-pink flowers; foliage emerges purplish-red and then turns green-bronze; persistent maroon fruit; narrow, upright form (6 x 4 m).

Chrysanthemums

As late as the 1970s, beds of chrysanthemums bloomed every fall at the Forestry Farm Park. Many of these had been bred by Les Kerr. In a talk on his work involving the propagation and over-wintering of chrysanthemums at the 1966 WSCH meeting, Kerr recommended that foliage be left on mums (as well as most other herbaceous perennials), until growth began in the spring. "Anything that can be done to hold snow and retard growth in the spring is desirable."

He introduced four chrysanthemums from the Sutherland Station: 'Sutherland Pink' (pink, pre-1959), 'Cree' (reddish, 1957), 'Early Autumn' (bronze-gold, 1958), and 'Popeye.' None are available today.

Other Introductions

While at Morden, Kerr crossed the flowering plum *(Prunus triloba)* with a Chinese species *(P. pedunculata)*. He introduced the resulting **'Prairie Almond'** in 1947 and several others in the 1950s. It had a graceful form and was covered in pink semi-double flowers before the leaves emerged.

Although widely distributed to members of the Garden Guild, University of Saskatchewan, in 1953, none are now available.

'Sutherland' caragana (1944)—vigorous, dense, and upright of habit; valuable as a specimen plant, tall hedge, or stock on which to topwork 'Pendula' or other ornamental types; 4 m in height; it blooms but does not set seed, thus is suitable for specimen or border plantings.

'Plume' caragana (1966)—fine textured with leaves resembling asparagus; light masses of foliage; arching and pendulous stems; 1.6 m; entered in the WCSH Merit Trials 1971.

'Sutherland' larch (1946)—dense dwarf but upright selection of Siberian larch; 4 x 2 m; no longer available.

'Wheeler' poplar (1944)—selected from poplar seedlings established by Seager Wheeler, Rosthern; vigorous male tree suited to shelterbelts; dense growth; dark green foliage. Now superseded.

'Sutherland Golden' elder (1977)—finely dissected, bright golden leaves; upright rounded form (3 x 3 m) with arching branches; small clusters of showy red berries; 1977 Award of Merit from Holland for the "best golden elder introduced into Europe"; 1979 WCSH Award of Merit. Readily available.

'Goldenlocks' elder—dwarf European red elder with very finely cut golden foliage; rounded form (1 x 1 m); resistant to sunscald; 1972 WCSH Merit Award. Available from a few nurseries, it is a hardy substitute for a dwarf Japanese maple and should be much more widely planted.

Willows

Les Kerr introduced three willows. 'Blue Fox' and 'Polar Bear' remain readily available.

'Blue Fox' *(Salix brachycarpa)*—1 x 1 m; globe-shape; blue-grey foliage.

'Polar Bear'—selection of *Salix silicicola* from the Athabasca sand dunes; silver-grey foliage; upright form (3–4 x 2 m).

'Fangstadt'—(1963) weeping form of sandbar willow *(Salix interior)*; 3 m; no longer available.

communication could be slow and difficult. But once you had his stamp of approval, he was the greatest of friends."

Much of the research and extension that he carried out as part of his job at the Forest Station benefited the nursery trades and in many ways was simply large-scale nursery management: propagation, weed control, pest control, nursery machinery, irrigation, and storage. He shared the results of his work freely with the nurserymen.

Until the 1950s, commercial nurseries used a costly heeling-in process in the fall, and this caused a bottleneck in the short spring planting season. The 1953 WCSH Proceedings noted that the experimental storage houses developed at the Sutherland Forest Station had been a success and predicted they would help solve this problem.

He was a driving force in the initiation of the Saskatchewan Nursery Trades Association (SNTA). In 1957, Les Kerr began to lobby some of the nurserymen in Saskatchewan to get organized into an association.

He invited nurserymen and horticulturists from the three prairie provinces to a two-day meeting at the Forest Nursery Station that ended in the formation of an association of nurserymen. Membership fees were five dollars per year, and there was a grand total of thirty dollars in the kitty when the meeting adjourned. (Krahn, 1957, 1977)

In 1967, Kerr was presented with a Savage rifle and case at the WCSH meeting in Banff in recognition of his service to the horticultural industry in the prairie provinces.

Plant Breeding

Although much of Kerr's initial horticultural breeding work took place in his early years at Morden, the majority of the plants that he eventually selected came out of his work at the Sutherland Station.

In terms of plant breeding and selection, he went well beyond his official roles, to the extent that in the last decade of his service in Sutherland there were hints that his federal supervisors were less than happy with his continued work with ornamental and fruit trees.

As Peters aptly wrote in 1988: "He has always been very willing to share the progeny of his plant breeding with other plant breeders and the effects of his imaginative and resourceful work will continue to show up in the years ahead."

Fruit

To give an idea of the consistent magnitude of his work as the research officer in charge of fruit breeding at Morden, in 1934 he reported that 66,462 fruit tree blossoms had been pollinated with pollen from a known source.

In a speech circa 1944, he noted his main interest since his return to Canada had been with fruit, "and we hope to have a very extensive demonstration along this line." To his largely urban audience, he added, "The Sutherland station is there to help you as well as the farmers and you are invited to make use of it." In 1999, George Krahn concurred that experimenting with fruit was "always high on his agenda."

Dr. Seager Wheeler and Les Kerr admiring Les's hybrid Prunus triloba *var.* multiplex *x* Prunus pedunculata, *c. 1940s. Courtesy Rich Krahn.*

He worked with plum, peach, apricot, cherry, apple, and pear, both in the greenhouse and outdoors.

Like the other horticulturists of his era, his correspondence and exchange of plant material was far-flung. While at Sutherland, he mentioned cooperation with Morden, the Minnesota Fruit Breeding Station, the South Dakota Agricultural College, the Central Experimental Farm, Vineland Experimental Station, Summerland, Mitchurin Institute in the USSR, the Ontario Agricultural College, as well as individuals such as Seager Wheeler, William Boughen of Valley River, Frank Skinner, and George Chipman.

By 1953, he had eighty acres of fruit seedlings at the Sutherland Station, many brought with him from Morden, including hybrids of Mongolian cherry, Nanking cherry, pear, and others.

Superintendent's residence, Forestry Farm, c. 1961. Courtesy Friends of the Forestry Farm House.

Sources

Blair, Kellet, and Kerr, *Budding and Grafting of Fruit Trees*, Dominion of Canada, Department of Agriculture, 1947.

Dabbs, Don, convocation speech, University of Saskatchewan, Spring 1982.

Dominion Forest Nursery Stations, Progress Report, 1947–1952.

Friends of the Devonian Garden, Publication # 21, *Edible Apples in Prairie Canada*, September 1991.

Kerr, W. L., speech, c. 1944.

———, letter to D. R. Robinson, 1 Apr. 1953.

———, "Pollination Studies with the Yuksa Sandcherry Hybrid," proceedings of the American Society of Horticultural Science.

———, Tree Distribution Report, PFRA Tree Nursery Station, Sutherland, 1965-66.

Krahn, George, Saskatchewan Nursery Trades Association, *Landscape Saskatchewan, 1957–1982*, 1986.

———, "Les Kerr, An Unforgettable Character," *Gardener for the Prairies*, Fall 1999.

———, Progress Report on the Organization of the Saskatchewan Nurserymen's Association, 6–7 Aug. 1957.

———, *Landscape*, Sept. 1977.

Melberg, Karin, *The Commentator*, 3 June 1981.

On Campus, May 1983.

Peters, P. J., *A Century of Horticulture in Manitoba, 1880–1980*, 1988.

Prairie Garden: 1959, 1963, 1965, 1970, 1979, 1980.

Rosthern Historical Society, *Old & New Furrows, The Story of Rosthern*.

Saskatoon *Star-Phoenix*, 4 June 1966.

Western Canadian Society for Horticulture Proceedings: 1948, 1950, 1951, 1953, 1958, 1960, 1964, 1966, 1971, 1971, 1979, 1984.

The Western Producer, 6 Aug. 1998.

Wilde, Douglas E., "The Evacuee," Sept. 2003.

Williams, Sara, "A Horticultural History of the Sutherland Forest Nursery Station," Friends of the Forestry Farm House, Inc., Saskatoon, Sask., 1998.

Wright, Percy H., Nanking *Besseyi* Hybrids, *Horticulture*, Feb. 1972.

———, "The Western Canadian Society for Horticulture," undated.

———, New Flowering Almonds, *Horticulture*, Feb. 1969.

www.littletreenursery.com

About the Author

With a passion for people, perennials, and puns, Sara Williams gardens on five acres of sand near Saskatoon. She is the former horticulture specialist with the Extension Division, University of Saskatchewan. She is also the co-author of *Best Trees and Shrubs for the Prairies* and *Perennials for the Prairies*, author of the award-winning *Creating the Prairie Xeriscape*, and a respected instructor of gardening workshops across western Canada and the Great Plains of the United States. *In a Cold Land* allows her to combine her background in English and history with her degrees in horticulture in sharing her interest in these horticultural pioneers of Saskatchewan.